"The next decade's leading corporations will be [illegible] to integrate quality principles into product development [illegible] environments. The authors give compelling discussions of the benefits and clear descriptions of the methods—both founded in AT&T experiences—for achieving quality product development capabilities which can make big differences."

H. Barry Bebb, Ph.D.
Barry Bebb & Associates

"For ten years I had responsibility for some 800 professional and support people with the mission to conceptualize, develop and manufacture weapon systems for the Army. I found this book saying precisely the things that I wish I could have done back then. We had all the problems and more alluded to here—compartmentalization of job functions, lack of focus, confrontations, low productivity. Such problems were characteristic of our labs at that time. Such a book as this would have been a golden resource for charting a way to improve product and productivity."

Dr. Royce Beckett
Mechanical Engineering Department, Auburn University

"I believe its unique feature is that the book *itself* is a roadmap—a step-by-step (or chapter-by-chapter) recipe for introducing and institutionalizing quality into product development and manufacturing environments."

Dr. Philip H. Francis
Vice President, Schneider North America

"This book integrates some very timely material on explicitly and implicitly introducing quality into the product realization process . . . This book will appeal to a large audience of cross-functional people concerned with product realization. This includes marketing, engineering, manufacturing, materials, purchasing, and human resources management people as well as those explicitly (or historically!) concerned with quality."

Dr. Steve Rohde, Manager
Systems Analysis & Information Management,
GM Systems Engineering Center

"I found the material extremely good—offering insights which I wanted to write down, to retain, and implement myself."

Dr. Philip Barkan
Department of Mechanical Engineering, Stanford University

Winning with Quality

Engineering Process Improvement Series

Consulting Editor, John W. Wesner, Ph.D., P.E.

Process management and improvement This includes books that take larger views of the field, including major processes and the end-to-end process for new product development. Books in this category include:

Changing the Way We Change: Applying Key Principles for Competitive Advantage
Jeanenne LaMarsh

Four Days with Dr. Deming: A Strategy for Modern Methods of Management
William J. Latzko and David M. Saunders

Winning with Quality: Applying Quality Principles in Product Development
John W. Wesner, Jeffrey M. Hiatt, David C. Trimble

Improving Functional Processes These are the specific functional processes that are combined to form the more inclusive processes covered in the first category. Books in this category include:

World-Class Systems Engineering
James H. Brill and Jerome G. Lake

Special process topics and tools These are methods and techniques that are used in support of improving the various processes covered in the first two categories. Books in this category include:

Engineering Methods for Robust Product Design: Advanced Taguchi Methods
C.M. Creveling and William Fowlkes

The Malcolm Baldrige National Quality Award: A Yardstick for Quality Growth
Maureen S. Heaphy and Gregory F. Gruska

Quality Function Deployment: How to Make It Work
Lou Cohen

Winning with Quality

Applying Quality Principles in Product Development

John W. Wesner
Jeffrey M. Hiatt
David C. Trimble

Addison-Wesley Publishing Company
Reading, Massachusetts Menlo Park, California New York
Don Mills, Ontario Wokingham, England Amsterdam Bonn
Sydney Singapore Tokyo Madrid San Juan
Paris Seoul Milan Mexico City Taipei

The publisher offers discounts on this book when ordered in quantity for special sales.

For more information, please contact:

Corporate & Professional Publishing Group
Addison-Wesley Publishing Company
One Jacob Way
Reading, Massachusetts 01867

Library of Congress Cataloging-in-Publication Data

Wesner, John W.

Winning with quality : applying quality principles in product development / John W. Wesner, Jeff Hiatt, David Trimble

p. cm.

ISBN 0-201-63347-7

1. Quality control. 2. New products. I. Hiatt, Jeffrey. II. Trimble, David. III. Title.
TS156.W47 1994
658.5′62—dc20 94-17265
CIP

0-201-63347-7

1 2 3 4 5 6 7 8 9—CRW—969594

First Printing, September 1994

To Monica, Mary, and Jessica, for your patience and support.

Engineering Process Improvement Series

Consulting Editor, John W. Wesner, Ph.D., P.E.

Global competitiveness is of paramount concern to the engineering community worldwide. As customers demand ever-higher levels of quality in their products and services, engineers must keep pace by continually improving their processes. For decades, American business and industry have focused their quality efforts on their end products rather than on the processes used in the day-to-day operations that create these products and services. Experts across the country now agree that focusing on continuous improvements of the core business and engineering processes within an organization will lead to the most meaningful, long-term improvements and production of the highest-quality products.

Whether your title is researcher, designer, developer, manufacturer, quality or business manager, process engineer, student, or coach, you are responsible for finding innovative, practical ways to improve your processes and products in order to be successful and remain world-class competitive. The **Engineering Process Improvement Series** takes you beyond the ideas and theories, focusing in on the practical information you can apply to your job for both short-term and long-term results. These publications offer current tools and methods and useful how-to advice. This advice comes from the top names in the field; each book is both written and reviewed by the leaders themselves, and each book has earned the stamp of approval of the series consulting editor, John W. Wesner.

Key innovations by industry leaders in process improvement include work in benchmarking, concurrent engineering, robust design, customer-to-customer cycles, process management, and engineering design. Books in this series will discuss these vital issues in ways that help engineers of all levels of experience become more productive and increase quality significantly.

All of the books in the series share a unique graphic cover design. Viewing the graphic blocks descending, you see random pieces coming together to build a solid structure, signifying the ongoing effort to improve processes and produce quality products most satisfying to the customer. If you view the graphic blocks moving upward, you see them breaking through barriers—just as engineers and companies today must break through traditional, defining roles to operate more effectively with concurrent systems. Our mission for this series is to provide the tools, methods, and practical examples to help you hurdle the obstacles, so that you can perform simultaneous engineering and be successful at process and product improvement.

The series is divided into three categories:

Process management and improvement This includes books that take larger views of the field, including major processes and the end-to-end process for new product development.

Improving Functional Processes These are the specific functional processes that are combined to form the more inclusive processes covered in the first category.

Special process topics and tools These are methods and techniques that are used in support of improving the various processes covered in the first two categories.

Contents

Foreword by Philip H. Francis *xv*
Preface *xvii*
Acknowledgments *xix*

Part I
INTRODUCTION *1*

1 What This Book Is About, and How It Can Help You *3*

2 Understanding Where You Are and Where You Want to Be *9*
Understanding Where You Are *9*
Understanding Where You Want to Go *13*
Deciding How to Proceed *13*

Part II
QUALITY CONCEPTS, PRINCIPLES, METHODS, AND TOOLS *15*

3 Overview of Quality Concepts *17*
Introduction *17*
Strategic Quality Ideals *20*
Organizational Enablers *25*

4 Improving Work Processes *33*
Background of Process Management *35*
Process Basics *38*
Continuous Process Improvement Model *40*

5 Process Re-Engineering *61*

Introduction *61*
Principles and Process of Effective Re-Engineering *66*
Step 1. Concept Development and Feasibility Analysis *67*
Step 2. Planning and Definition *72*
Step 3. Development *74*
Step 4. Verification *76*
Step 5. Deployment *77*

6 Principle-Based Problem Solving *81*

A Rationale for Effective Problem Solving *81*
The Five Principles Behind Effective Problem Solving *83*
Clearly Defined Problems *84*
Participation Without Criticism *84*
Addressing Root Causes, Not Symptoms *86*
Prioritization of Causes and Solutions *92*
Data-Driven Decisions *94*
Some Comments on Tools *95*
A Problem-Solving Process *96*

7 Concurrent Engineering in a Process-Centered Environment *105*

A CE Story *105*
Critical Success Factors for Concurrent Engineering *113*

8 Role of Teams *135*

Teams *135*
The Team Environment *142*
Self-Directed Teams *155*
Putting a Team to Work *164*

9 Goal Setting for Product Development *169*

Introduction *169*
Customers' Needs *171*
Performance Level Against the Competition *179*
Business Objectives *182*
The Vision for Product Development *184*
Establishing Goals for Product Development *187*

Metrics 187
Deploy These Directions and Goals Throughout the Organization 188

Part III
DEPLOYING QUALITY PROGRAMS AND PRACTICE 191

10 Developing Your Program 193
It's Your Turn 193
Deciding Where You Want to Be 195
Understanding Where You Are 197
Planning How to Proceed 199

11 Lining Up Resources and Building the Support Structure 207
Obtaining Funding 208
Gathering the Team 221
The Support Structure 224

12 Getting Started 231
Recapitulation 231
The Preliminaries 231
Starting Your Actual Quality Work 238
Looking at the Big Picture 244
Summary 248

13 Total Quality Management 249
Total Quality Management 249
Building the Infrastructure 255

Part IV
LEARNING FROM YOUR EXPERIENCE 257

14 Celebrating Your Results 259
Introduction 259
Using Rewards to Drive Desired Behavior 261
Special Compensation 262
Awards 263
Celebrations 267
Mementos 268
Other Activities 269

Special Rewards for the Quality/Process Team *269*
Summary *269*

Epilogue *271*

Appendix *273*

Glossary *277*

References *289*

Index *295*

Foreword

We are all aware of being caught up in a new industrial revolution, marked by keen worldwide competition. Unlike manufacturing as practiced before about 1960, the United States alone no longer dominates the world market of new product development and manufacturing. To succeed in this new world, all companies—small and large alike—must understand this new reality and face it by continuously improving the speed and precision of their internal systems. Manufacturing firms that will win in the future must do an outstanding job at product realization: the rapid conception, development, and production of products that the marketplace welcomes.

Change today is pervasive, with new terminology and approaches, new tools and practices. It is ever more difficult to keep up and operate within this whirlwind of change. Because of this everyone speaks in the shorthand of buzz words. Too often, however, mastery of clichés and acronyms is confused with mastery of the subject matter itself. Winners realize this, and dig deeper. Pretenders are content merely with the babble of buzz. It is vitally important that people engaged in the product realization chain learn and apply proven methods for improved practices. Books like this one do us all a service by taking us beyond the catch phrases of the day, to show us the way to improved performance.

Those who are intent on learning how to improve their development engineering processes systemically will find this book's approach refreshing. The authors, all experienced engineers at AT&T Bell Laboratories, have suffered through the difficult experience of learning how to transform their product development processes from the way development engineering used to be conducted, to the way it must be performed in the future. They have walked the talk. Their journey was successful. In the process of their journey, they have learned a great deal about what works and what doesn't. This book synthesizes their collective experience in a readable form that teaches the reader not only the *why*, but also the *how*. They translate the tenets of modern quality into prescriptions for success in building effective product development processes. In so doing, they have done a real service to engineers and managers who are smart enough to know they need to work smarter.

They do this by borrowing key precepts from the field of quality, which has developed and matured in recent years and now plays a key role in improving all aspects of organizational competitiveness. In this book, the authors apply these precepts to industrial product development—or more specifically to the functions of concurrent engineering. The reader will learn quality principles, methods, and tools, and how they can be applied to strengthen concurrent engineering functions. The authors' point of view is that, just as in manufacturing, the product development *process* is the key to serving up better products to the marketplace. The many available tools are the enablers, but the process—or, as the authors call it, the *process management*—is what makes it work. This viewpoint is both correct and profound.

The authors have succeeded in producing a book that is practical, yet doesn't sacrifice depth and insight. It is prescriptive, and can be used effectively by development engineering groups—in small and large companies alike—to improve their processes and leap ahead toward *true* world-class performance. It is easy to read; in fact, it is even entertaining. It deals extensively with human issues, for people are the developers, the owners, and the operators of the new process systems. It emphasizes the crucial aspects of leadership, employee empowerment, individual responsibility, and accountability. In addition to being easy to use, there are several levels to this book; it will appeal to the newcomer to product development systems, to the new manager, and to the experienced engineer and engineering manager alike.

Philip H. Francis
Vice President—Technology
Schneider North America

Preface

You can change the way your company thinks about product development. Better than that, you can be the change agent for breakthrough improvements in your company's performance. In this book, we take the first steps with you. We begin a quest for excellence in product development, including faster time-to-market, higher-quality products, lower overall costs, and more satisfied customers.

We have worked with teams to implement programs in our product development organizations that have reduced product development intervals more than 50 percent, and have decreased rework from a typical four cycles for development to nearly one. The product-design defect rate as measured through change activity improved threefold in less than two years, and the number of *met* commitments to internal customers increased from 20 percent to more than 80 percent.

How did we achieve these results? By focusing on the *processes* and *people*. By instituting an environment for quality improvement and process management that goes beyond slogans and short-term programs. By following the principle that you will never achieve any more than you plan, and that the plan will be no better than the vision that you have for your business.

This book takes you on a journey—the same journey we took to achieve these gains. We begin with the groundwork laid by others, and add to that our own experiences in improving our product development processes. You pick the destination, you provide the power, and we provide a tool kit and framework to support breakthrough improvements in *your* product development process.

Acknowledgments

We authors wish to thank many of our colleagues, who participated with us in the journey described in this book. Special thanks go to Herb Burton, mentor and coach to all three of us, who promoted, supported, and championed the work we have described, and continually encouraged us.

Carol Carlton and Fran Owitz, who helped us in countless ways!

Andrea Rhodes and Joe Serhan, managers at AT&T's Denver factory, who were our partners in bringing the R&D and manufacturing engineering organizations together, and helped make our individual processes for new product introduction into a single, seamless process. Joe's entire department played a significant part in making things work.

Colleagues throughout AT&T who helped us find references, gather data, and fill in gaps, including Dave Gatenby, Dave Gower, Bill Harrod, Sarah Lincoln-Thiele, and Dan Stanzione. Dave Gatenby and Lou Enloe served as internal reviewers.

In addition to our work colleagues, we must thank the people who reviewed our manuscript, for the guidance they gave us to make the book more focused and tuned to the needs of our customer: you, the reader. Our reviewer thanks go to: Phil Barkan (Stanford University), Barry Bebb (Barry Bebb & Associates), Royce Beckett (Auburn University), Herb Burton (AT&T Bell Labs), Don Clausing (Massachusetts Institute of Technology), John Eimermacher (University of Dayton), Phil Francis (Schneider North America), Richard Hirsch (AAI Corporation), Glen Johnson (University of Dayton), Steve Rohde (General Motors), Warren Seering (Massachusetts Institute of Technology),

David Ullman (Oregon State University), and David Zenger (Worcester Polytechnic Institute). Our editor, Jennifer Joss, and her assistant, Danielle DesMarais, also helped us materially to fine-tune the manuscript, and kept us energized to get done on time.

Thanks, also, to our families, who (mostly cheerfully) agreed to let us out of some family obligations so that we would have time to keep writing.

Part I
INTRODUCTION

1 What This Book Is About, and How It Can Help You

WELCOME ABOARD! This is your guidebook for a journey toward productivity improvement in a product development environment. It is based on our own quest, in support of our product development work at AT&T Bell Laboratories, to shorten development intervals and improve product quality, while at the same time lowering costs.

Some of you may read "Bell Laboratories" and say to yourselves, "This is probably too esoteric for me—those folks do all that fancy *research*." The truth is that most of the people of AT&T Bell Laboratories develop new hardware and software products and services. Some Bell Labs organizations do conduct research for the communications systems of tomorrow, but in terms of numbers of technical people and dollars spent, there is much more "D" than "R." The experiences conveyed in this book come from many years of developing customer premises products—apparatus like telephones, personal computers and small-business telephone systems, which are manufactured in substantial quantities.

Our means to achieve productivity improvement and, along with it, product improvement, has been to adapt quality principles used in manufacturing and service industries to the development environment. This book presents a number of key concepts and practices, plus a description of how we applied them. Examples of programs that worked, and lessons from some that did not, are also included.

We have found that quality improvement is not something that you do once and then sit back to enjoy the results. Quality improvement is a *journey*. It is, in fact, a cultural journey, from focusing on *What* to do, through focusing on *How* to do it, to finally focusing on the people *Who* are doing the job.

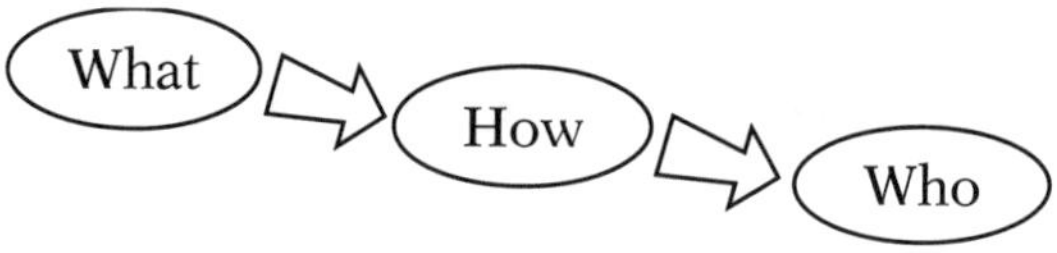

Our traditional product development culture was focused on *the product*—the *What*. Operating in this mode, we did whatever was necessary to get the product out: test, rework, unexpected overtime, or whatever "heroic acts" were necessary: producing product was the only goal.

Our change in focus to *the process*—the *How*—began during the 1980s. We turned to managing our processes and systems in such a way that we met or exceeded the expectations of our customers. We actually achieved higher product quality at lower cost, by improving our processes.

When we first began to apply what we were reading in the literature about quality and process improvement, we were frustrated. We did not seem to be collecting data at a rate sufficient to permit us to realize the optimistic gains described in the literature. We could not see, from the data we were collecting, opportunities to improve our processes that we could apply, and then, in a relatively short time, measure improved results. It took us quite a while to understand that we were not simply missing the point or doing things wrong. The difficulty was that much of the early work in process management and improvement was done in manufacturing or administrative environments, where the cycle times are sufficiently short that one could indeed observe a problem in the data, devise a process improvement, put the improvement into place, and observe the improved results in minutes, hours, or, at worst, a few days. We were looking for the same sort of performance in our development environment—where the cycle times (especially before we undertook our process improvement activities) were on the order of months or years, not hours or days. We could not hope to see the results of our improvement activities in the same short times about which we were reading. Once we understood what we were facing we were able to adjust our expectations and our approach, and make the concepts work for development.

Having embraced this concept, we came to the realization that if the *How* is the key to improving the *What*, then the people—the *Who*—are the key to improving the *How*. In essence, the people are central to improving the *whole* business! People, especially people working in teams, are the key to achieving success in productivity improvement. While this may seem obvious to some, the reality is that the environments of today have not universally accepted that it is the people who are the clear route to quality success.

We are without question talking about something big here—not just a little change in how you go about your day-to-day business. Working "a little harder each day" won't achieve the kinds of results about which we are talking, nor will it suffice simply to enjoin people with slogans, as we did in the past, and then go on with business as usual. We are talking about an approach to working a little "smarter" each day. It involves a major revamping of how everyone in the organization views their work. It is indeed a *paradigm shift*. But it can be done! We have seen it work. One of our teams cut product realization intervals in half and reduced rework from a typical four cycles per development to, essentially, a single cycle. They reduced product design defects, as measured by the number of design changes, by a factor of three in less than two years. At the same time they improved the number of *met* commitments to internal customers from 20 percent to more than 80 percent. Another team working on a related product achieved an *order of magnitude* improvement in software quality in a period of four years.

We've also tried some potential "improvements" that did not work very well, so we had to regroup and try a different approach. Overall, we achieved *measurable* success, and that is what we want to share with you.

What You Can Expect to Gain from this Book

We do not provide "six easy steps to instant success"—or even to short-term success. As with anything that is really worthwhile, the quality improvement journey is arduous. If you are willing to commit to the journey, we provide in the pages that follow:

- A comprehensive presentation of the key quality concepts, principles, and practices used by world-class companies today
- A basic set of tools and techniques that are being used successfully by people working on goal setting, process management and improvement, problem solving, and team building

- An approach for success, derived from our experiences, that you can adapt to your own situation
- Examples from our own experience, and from the experiences of others, to support the approach
- Some warnings about traps and pitfalls we encountered, with suggestions on how to avoid them

This book is organized into five parts, beginning with this Introduction and ending with a collection of references to help you get the most from the core of the book. The core is contained in the central three parts:

- Quality Concepts, Principles, Methods, and Tools
- Deploying Quality Programs and Practice
- Learning from Your Experience

To make the book more readable, we have included many examples and figures.

There are two ways to use the book:

- Read it from cover to cover, or
- Read the "minimum subset" to get a feel for what is there, and then read the specific sections you need to do your job.

It is not necessary to read the book from cover to cover to get something from it. It depends upon your needs and your level of experience with quality and with process management and improvement. We do suggest that you read Chapter 2 and the **Overview** in Chapter 3. These will tell you what to expect from the book, and give you an idea of how to find what you want in the rest of the book. If you are just getting started with quality and process, you may want to read straight on through to the end. If you are already deeply involved in making the change to a quality-focused environment, and seeking help with some specific problems, you may prefer to turn to a particular chapter.

Part II is a compendium of resources: concepts, principles, methods, and tools. It begins in Chapter 3 with the overview of quality concepts and principles mentioned above. Chapter 4 presents a comprehensive view of process management concepts and applications. This is a *doers* guide for making process management work in your organization. Chapter 5 addresses those situations where major surgery is needed to reach a stable new state—where "re-engineering," rather than incremental improvement, is needed.

Chapter 6 describes several tools we have used for solving quality and process problems, and Chapter 7 considers the important concept of *Concurrent Engineering*.

Chapter 8 talks about *teams*—teams of various kinds to fit different needs. We extend this discussion to include the powerful concept of *empowered teams*. Teams used in a concurrent engineering environment can by themselves lead to significant improvements in development and manufacturing.

Chapter 9 provides a specific methodology for establishing and deploying directions and goals for a development organization. Development managers and coaches—those staff who end up doing a lot of the work for their top-management leaders—will find this information valuable for bringing customers, the competition, and even the overall business directions into the goal-setting process for the development organization.

In Part III we change gears, from describing principles, methods, and tools to describing how to implement a comprehensive quality program. Chapter 10 takes you through the experience of developing your own program to establish an environment focused on quality and on process management and improvement. Chapter 11 addresses lining up the resources needed to make your program happen, and building the necessary support structure. Like any journey, the quality journey cannot be undertaken without adequate resources, including funding, management support, and people to do the work.

Chapter 12 tackles actually getting started. The work has to be divided among the participants, and you need to select tools. We describe some ways to help get the new work off the ground, and to support the people working on the changes as they come up to speed. This leads right into Chapter 13, which addresses managing the overall quality system—Total Quality Management (TQM). Chapters 12 and 13 are especially important; they address the cultural integration that must take place in order for your quality and process work to succeed.

Part IV looks at the ongoing journey. Chapter 14 discusses the importance of clearly identifying what you have learned from your experiences, and your role in sharing your organization's successes and near misses. Finally, the epilogue looks to the future. The theme is *continuous improvement*, because completing one cycle of the journey does not mean that you are "done." Your competition is probably embarked on a similar journey, and only continuous improvement will keep you where you want to be.

Throughout the book are examples from our own experience and the experiences of others. To make it easy to find these examples, most are set off from the surrounding text by being indented from both margins. Many start off with the heading, **Example**. Some are identified more specifically, to correlate with the accompanying text.

The book concludes with an appendix, a glossary of terms, a list of references, and a thorough Index.

2 Understanding Where You Are and Where You Want to Be

We appreciate that reading a how-to book like this, with the intention of applying what you learn, entails a substantial commitment of time. In today's work environment, we all find ourselves rushed to meet the pressing day-to-day needs of the business. As associates in a development organization in a large company, we certainly appreciate the value of time, and the seeming lack of it. Before you commit yourself to this journey, you may want to assess where you are today, and consider where you would like to be. Once you have this initial insight, you will be in a better position to decide how to proceed.

UNDERSTANDING WHERE YOU ARE

You can evaluate your organization's present quality levels by making both external and internal assessments:

- External Assessment: Measure the satisfaction level of your customers (both internal and external), and your performance level against your competitors.
- Internal Assessment: Understand your current levels of implementation of the basic quality principles.

External assessment is the best snapshot of your current situation. In Part II of this book we present techniques to conduct an external study of customers and competitors. Since you may not be sure if you want to read this entire book from cover to cover, or you may want to select just a few chapters to review, we have provided a *Five-Minute Self-Assessment* in Figure 2.1. You can get a quick scan of your situation from this *internal* assessment.

Take time now to do the Five-Minute Self-Assessment in Figure 2.1

Five-Minute Self-Assessment	
Yes__ No__	We *know who our customers are* (both internal and external).
Yes__ No__	We talk with our customers to *understand their needs.*
Yes__ No__	We understand what is *most important* to our customers.
Yes__ No__	We understand how our customers feel about *our performance.*
Yes__ No__	We *measure* our performance in meeting customer needs.
Yes__ No__	Our customers are satisfied or even *delighted* in our performance.
Yes__ No__	We know who our *competitors* are.
Yes__ No__	We *compare our performance level* with the best of our *competitors.*
Yes__ No__	We *compare our performance levels* with *best-in-class companies.*
Yes__ No__	Our current performance is equal to or *better than our competitors.*
Yes__ No__	We have *clearly defined processes* for our work.
Yes__ No__	Our processes are *documented.*
Yes__ No__	We have *metrics* for each process, and continuously measure process performance.
Yes__ No__	We use *data,* not intuition, to drive decision making.
Yes__ No__	We set *target levels for our process metrics.*
Yes__ No__	We *continuously improve* our processes based upon performance data collected.
Yes__ No__	We have well-defined *corrective-action processes* in place.
Yes__ No__	We have *identified key goals* based upon our customers, our competitors, and the needs of our business.
Yes__ No__	We have clearly *communicated these goals* to our associates.
Yes__ No__	We know the *benchmarks* for these goals, and use these benchmarks for setting objectives.
Yes__ No__	Our *associates align with* and support the objectives.
Yes__ No__	We make effective use of *teams* to execute processes.
Yes__ No__	Our associates feel *empowered* to make decisions, and are actively engaged in the day-to-day improvement of our work processes.
Yes__ No__	Our *value system, performance review process, and reward structure* support continuous improvement of our processes.

FIGURE 2.1 *Five-Minute Self-Assessment*

If you answered *No* to any of these questions, then this book has something to offer you. If you had only a few *No*s, then you may want to be selective in the chapters you read. You can use the table of contents to direct you to a specific topic area, or you can review the latter part of Chapter 1, where each chapter is summarized. If, on the other hand, you answered most of the questions *No*, you probably have more work to do and could benefit from reading the book from cover to cover.

If you passed our short test with flying colors, you may want to try a more in-depth assessment. The criteria for the **Malcolm Baldrige National Quality Award** provide a comprehensive vehicle for self-assessment. The Baldrige Award criteria, which have changed a little each year (becoming ever more challenging—essentially "raising the bar"), are grouped into seven categories, with point scores assigned to each, as shown in Table 2.1.

Because the Baldrige Award criteria are lengthy and are updated annually, we do not include them here. You can obtain a current copy of the Baldrige Award Criteria from the United States Department of Commerce,[1] who manage the Award Program. Additional information about the Baldrige Award, including a *Self-Assessment Tool* (Brown 1993, 82-87) can be obtained from the American Society for Quality Control,[2] who administer the Award for the Department of Commerce. Use these criteria to analyze your organization. As an R&D or product development leader, you will probably not be able to address all seven sections fully, since some relate to the entire business. (You may want to convince someone to assess the entire business to the Baldrige criteria, which will help you with *your* assessment—and may very well help the company as a whole!)

TABLE 2.1 *Baldrige Award Criteria*

Category	Points
Leadership	100
Information and analysis	60
Strategic quality planning	90
Human resource utilization	150
Quality assurance of products and services	150
Quality results	150
Customer satisfaction	300
Total points	1000

1. Individual copies of the award criteria can be obtained free of charge from: Malcolm Baldrige National Quality Award, National Institute of Standards and Technology, Route 270 and Quince Orchard Road, Administration Building, Room A537, Gaithersburg, MD 20899-0001.
2. American Society for Quality Control, Customer Service Department, P.O. Box 3066, Milwaukee, WI 53201-3066. For information phone 800-248-1946.

Here are three sample Baldrige questions:

- From the section on Leadership: "Describe the company's quality values, how they are projected in a consistent manner, and how adoption of the values throughout the company is assessed and reinforced."
- From the section on Human Resource Utilization: "Describe the means available for all employees to contribute effectively to the company's quality objectives; summarize trends in this involvement."
- From the section on Quality Results: "Summarize trends in quality improvement, based upon key measures of business processes, operations, and support services."

Your lowest-scoring areas may suggest where you need to focus your improvement efforts.

Another set of criteria that might be used for self-assessment are those for the **Shingo Prize for Excellence in Manufacturing**. Established in 1988, the Shingo Prize recognizes companies in the United States, Canada, and Mexico that demonstrate excellence in manufacturing leading to quality enhancement, productivity improvement, and customer satisfaction. The concept that underlies this prize is that world-class status can be achieved through lean, just-in-time manufacturing processes, systems, and methodologies. The criteria for the Shingo Prize are published annually in a *Guidelines* booklet.[1]

It is not so important *which* tool you use for self-assessment, so long as you are able to identify areas for potential improvement, and also areas of strength. You may later find it useful to re-assess yourself periodically (perhaps annually) with the same tool to see what progress you are making.

Although internal assessments are useful for beginning to understand where an organization is, external assessments are vital. Your customers' perceptions of your organization are a key to understanding your present level of performance. You also need to assess yourself against your competition—those people who sell products and services against yours. If you do not know your customers' perception of your performance versus that of your competitors, then this is an important area to examine. We specifically address this in Chapter 9. It is very difficult to understand where you are today without a strong external focus—on both your customers and your competition.

You may also want to assess yourself against others who, while they may not compete with you for business, use manufacturing and other processes similar

1. Contact the Shingo Prize Office: Shingo Price for Excellence in Manufacturing, College of Business, Utah State University, Logan, Utah 84322-3521.

to those used in your organization. Such "process competitors" may even exist in other parts of your own company, where they employ components and/or processes similar to those you use.

We return to this question of understanding where you are in more detail in Chapter 10.

UNDERSTANDING WHERE YOU WANT TO GO

Customer perceptions, benchmarking data, and self-assessment results will not only tell you where you are today, they can also help you develop an idea of where you would like to be. This is especially true if you have been able to obtain real competitive comparison data.

Does your chief competitor generally introduce new products in one-third the time you do, with fewer prototype cycles? Did your external customers tell you whose competitive products they almost bought (or did buy!) instead of yours? Do only a small fraction of your customers say that they will buy from you again without shopping around? Do your customers perceive your chief competitor's quality to be better than yours? All of these data can help you to set your goals for improvement.

If you seek a real challenge—if you really want to "go for the gold" of world-class quality—you can go even further and identify **Best Current Practices**—and then combine the best parts of several: the "best of the best". This represents what *can be achieved* if you could put everything together just right. If you could do this, you could end up being better than your best competitor today. Rest assured that your leading competitor is trying to do just that!

In Chapter 10 we discuss in detail how to get an understanding of *where you want to go*. We also include an example of how to come up with a "best-of-the-best" process.

DECIDING HOW TO PROCEED

The "gap" between your current situation ("where you are") and the picture you have built of where you want to be will help you decide how to proceed. Perhaps things are going fairly well: nothing is really out of control, but you are not satisfying yourself or your customers. You may want to implement only a few improvement programs. For this you may want to refer to specific sections of this book, perhaps using the index for guidance.

If, on the other hand,

- Your development intervals are significantly longer than those of your competitors, or
- Your development costs are significantly higher than those of your competitors, or
- Your customers have indicated that they are not very likely to buy from you again, or
- Your current processes are *seriously broken* or *seemingly non-existent,*

then you probably need to implement a comprehensive program.

Just how rigorous you have to be and how much work this means depends, of course, on the size of your company. The concepts and principles can be applied to advantage in *any* company. The thoroughness of your assessments and the complexity of your plans will vary greatly between a large corporation like AT&T and a small development and manufacturing company or an entrepreneurial "skunk works" activity of a larger company. It is important that you keep this in mind, and not attempt to set up a more complicated activity than your environment really needs. If what you propose seems to be too complex or too costly, you are liable to encounter more resistance than you can overcome.

Before we start addressing in detail the gaps and the "getting started" items, it is necessary to understand the methods and tools thoroughly. Chapters 3 through 9 address this need. Then in Chapter 10 we begin to apply all the concepts.

Part II

QUALITY CONCEPTS, PRINCIPLES, METHODS, AND TOOLS

3 Overview of Quality Concepts

INTRODUCTION

When we first began looking at product development productivity during the 1980s, we did not simply say to ourselves, "let's increase productivity," and then set off to work on the issue. We almost always were reacting to a problem (usually the most pressing problem of the month). The problems ranged from long development intervals to expensive design changes that impacted product already in the field. Our early efforts to work out these problems were unfocused, and often relied on one or two quality programs that were popular at the time. We would try one thing for a while, and when the interest in that died down we would try something else.

During these early efforts, we did not have a clear focus on what a quality system was, and whether or not quality programs were the answer to these very critical problems in development. We certainly read about many quality programs and their successes, but we also read about many failures (and heard of even more through the grapevine). We struggled to get a handle on the basic principles behind all of these quality programs to see if answers could be found for our problems.

We knew that other companies were implementing a variety of quality initiatives. The emergence of a world economic structure and global competition for goods and services were forcing many American businesses to change. Consumer electronics and the automotive industry are two good examples. Until recently, these areas were controlled by U.S. companies. Today, Japanese firms dominate many segments of these markets.

We have felt similar pressures. Since the break up of the Bell system in 1984, AT&T has struggled to find its feet in a competitive marketplace. For the past ten years we have worked to find the right solutions to make us competitive and profitable in this new environment. We have learned from industry leaders (including firms not in the electronics industry), books and journals, and from practical experience applying new ideas.

Lessons came from the Japanese as well. Interestingly enough, many of the principles employed by the Japanese were taught to them after World War II by people from AT&T (both Deming and Juran worked for Western Electric). Japanese success during the last twenty years has hinged on the quality of their products, and their ability to rapidly bring new products to market at higher quality levels and at lower costs than their competitors. In other words, manufacturing and R&D played a key role in this evolution for Japan.

Convinced that the practices used by Japanese firms and other industry leaders held the answers to our productivity problems, we went to work to learn what they were, and then began applying them to our development work. In this chapter we introduce the quality principles that we have found to be effective. We intentionally do not spend too much time on each principle in this chapter, since we want you to get to the heart of the book as quickly as possible. This chapter gives you just enough to get the gist of each area, and then we move on to develop them more fully in Chapters 4–9.

A Look at the Quality Ideals

W. Edwards Deming, Joseph M. Juran, Philip B. Crosby, Walter A. Shewhart, Kaoru Ishikawa, Genichi Taguchi, Armand V. Feigenbaum, and others have contributed to today's quality principles.[1] Companies such as Florida Power and Light, the first company outside Japan to win the Deming Prize, and Motorola, the winner of the first Malcolm Baldrige National Quality Award, pioneered the quality movement in America. Xerox also played a lead role as they used these quality techniques to turn their copier business around and regain market share from the Japanese. Xerox's direct tie of business success to the quality principles set a good example for other companies looking to increase the effectiveness of their business (Kearns and Nadler, 1992).

1. Several authors have written about the philosophies and contribution of these quality pioneers, including V. Daniel Hunt (Hunt 1992, 47) and Lloyd Dobyns and Clare Crawford-Mason (Dobyns and Crawford-Mason 1991, 52).

Our study of these pioneers' work in quality, as well as the implementation efforts of many companies, surfaced five basic quality ideals. We have found these ideals to be the underpinnings of today's quality programs. They are:

1. Clear Vision and Goals Communicated to All
2. Total Employee Involvement and Empowerment
3. Customers as the Key Drivers for Change
4. Focus on Process: the How of Doing Work
5. Metrics for Continual Performance Assessment

We termed these areas the *Strategic Quality Ideals* to give them a name that we can easily refer back to later. The reason for the word *Strategic* is simply that we found that implementation of one area without the others results in mediocre success. These principles are interrelated and therefore part of a strategy for improving product development performance. The word *Ideal* is used to indicate that these are ideal states. The word *Quality* could be included or not; we decided to include it.

With our new-found understanding of these quality ideals, we thought we had it made. Not so. We soon realized that making these ideals work for new product development is not easy. Complaints ranged from "No leadership," to "This is just another flash-in-the-pan program." Our history of one program after another was coming back to haunt us. These difficulties surfaced organizational qualities that must be present for successful implementation of the Strategic Quality Ideals. We refer to these qualities as *Organizational Enablers*. They include:

1. Leadership and Commitment at the Top
2. Support and Alignment among Middle Managers
3. High Morale and Satisfaction of Employees
4. Strong Values Shared by All
5. Reward Structure Linked to Contribution

A reasonable first step is to look briefly at the Strategic Quality Ideals and Organizational Enablers, and understand how the quality programs of today relate to these principles. In Chapters 4 through 9 we add detail to these principles and discuss our own experiences. At the beginning of each chapter, a checklist highlights the ideals and enablers that are developed and discussed in that chapter.

STRATEGIC QUALITY IDEALS

Clear Vision and Goals Communicated to All

Vision, mission, goals, objectives, directions, strategy, and plans: all are ways of articulating the future state and direction of an organization. Regardless of the words used, this Strategic Quality Ideal simply means that the short- and long-term intentions of the organization need to be clearly communicated to the doers in such a way that they can integrate these directions into their daily work. Successful integration of the vision and goals into daily work allows each individual decision to push the organization that much closer to its desired future.

The use of the word *vision* rather than *goal* in business has become more popular in the last ten to fifteen years. It conveys a sense of hope, long-term thinking, and something that is not concrete. A vision is something dreamed, imagined. Unlike a goal, which is something you expect to reach, a vision is something you strive for, but never quite reach. Visions instill hope and create a belief that a future is out there waiting to be realized. Visions form the basis for strategic intent.

Vision is commonly defined as a perception of a future state or condition through imagination or foresight. James A. Belasco, in his book *Teaching the Elephant to Dance*, formulates vision from a business perspective as "a statement of what you want your organization to be. It conveys a picture of where you want to go and how you want to get there. It is a simple-to-understand, inspirational, focusing statement" (Belasco 1990, 104).

Visions alone, however, do not constitute the Strategic Quality Ideal of Clear Vision and Goals Communicated to All. The strategy for moving forward must be established, long-term plans need to be developed, and measurable objectives and goals need to be deployed to everyone in the organization. This forms the basis for what we commonly call *Policy Deployment* today.

The newer terms *vision* and *directions* are very good complements of the traditional concepts of goal, strategy, and plan. But one set is not a substitute for the other. One provides a long-term, imaginative view, while the other supports the concrete, near-term work, and is measurable. Communicating a picture of the future state along with specific plans each step of the way is a fundamental element of the quality movement. Achievement of this ideal in an organization is evident in employees who are hopeful and excited about their future and the future of *their* company, and who have near-term goals and objectives that they believe in and want to achieve.

Total Employee Involvement and Empowerment

A tremendous resource for problem solving and quality improvement lies with employees. Think about who actually interacts with the customer, handles the product, answers the phone, operates the equipment, orders the material, pays the bills, and does the multitude of activities necessary to make the business function on a day-to-day basis. With these people, a vast knowledge pool exists, along with creativity and ingenuity that surprise many managers. Tapping their profound knowledge of day-to-day business processes and their creative energies is at the heart of this Strategic Quality Ideal.

The transition to team-based structures has been one mechanism that capitalizes on Total Employee Involvement, and brought to the surface the many possibilities of increased employee participation. Team approaches are so vital that many companies such as AT&T are including team skills as an important performance evaluation factor.

A second important element of this ideal is *empowerment*. Employees who feel that they are simply tools for a job will perform to this expectation. Employees who contribute to and have a say in the operations they perform develop a sense of ownership and pride in their work, and will continuously strive to improve and change—not because someone told them to, but because they want to do better.

To illustrate the relationship between the Strategic Quality Ideals and demonstrate the reason for including *strategic* in the name, consider the example of the first ideal discussed, Clear Vision and Goals Communicated to All, and how this interacts with the concept of empowerment. Consider what could happen in an organization without a uniformly accepted vision and plan that embraced and implemented this concept of empowerment. The lack of alignment from one group to another can result in confusion and misdirected activities: ". . . empowering individuals or units in an unaligned organization often exacerbates conflict, increases stress and undermines organizational performance" (Smith 1989, 83).

In the February 1992 issue of *Long Range Planning*, Colin Coulson-Thomas, Chairman of Adaptation LTD, discusses the results of a 1991 British Institute of Management survey that showed that of the fifty-nine surveyed organizations, 88 percent were creating slimmer and flatter organizations, 79 percent were undertaking more work in teams, and 78 percent were creating a more responsive network organization. "In such circumstances, involving change and uncertainty, a clear vision and strategy is essential. Without it organizations can become fragmented as devolution and delegation occurs during the transition from the bureaucratic to the emerging network organization" (Coulson-Thomas 1992, 82).

The quality ideals are, for the most part, a dependent set, and therefore strategic in nature in that they interact and are related to one another. Together they can produce a positive result, and independently they can make matters worse.

Customers as the Key Drivers for Change

Without customers, there would be no business to manage. Identifying the customer as the key driver for change is acknowledging that total customer satisfaction is paramount to business success. Moreover, customers' needs for most businesses are constantly changing. This ideal is reflected in many companies' objective to delight the customer, or keep customers for life.

In the textile industry, Milliken Company, an early Baldrige winner, actively involves customers. As related by a Milliken employee to us during a benchmarking visit to Milliken, the company was concerned over the slow growth in orders from Taco Bell for uniforms when they knew Taco Bell was growing rapidly. After involving Taco Bell on a Customer Action team, they discovered that the color offering from Milliken was not suitable for their needs. Taco Bell really needed uniforms that did not get dirty so quickly in the taco business—you can probably guess the reason. After doing a straight dye match from a taco sauce sample, Milliken now offers a color that meets the needs of this customer.

Whether the information is gathered from insightful anticipation, direct customer involvement, random surveys, or customer feedback, the message is the same. The customer must be the key driver for change in the products and services offered if the business is to survive and grow. We must become more involved than ever with customers, so that new products match and even exceed customer expectations.

Focus on Process: the *How* of Doing Work

Every job can be described as a process. Whether you build houses, make titanium metal, or develop new products, each can be described as a set of discrete steps that lead to the desired end. This change in orientation to process was one of the earliest applied quality principles, advocated strongly by Deming and others who helped teach Statistical Process Control (SPC) to the Japanese in the postwar period. The application of SPC in manufacturing was just the first of many tools centered on this ideal.

Today, process management and improvement tools are being applied to all business processes, not just those in manufacturing. Research and development, logistics, and customer service are a few of the areas in which process focus is changing the way companies look at their operations. As each primary business function is defined as a process, the customer(s) and outputs for that process, *even when it is an internal customer*, can be identified. At the same time, the inputs to the process can be identified with their suppliers. Process performance can be quantified or measured in terms of the quality of inputs from the suppliers, the quality of outputs to the customers, and the overall performance of the process itself (time, cost, etc.). The process can then be continuously improved to optimize these measures.

The result: more time invested improving the processes that develop, manufacture, and offer a product or service, and less on inspection or rework of the final product. This yields faster, more efficient operations that produce higher-quality goods and services the first time around, to the satisfaction of both the business managers and the end-customers.

Metrics for Continual Performance Assessment

Metrics determine your performance in all of the other Strategic Quality Ideals. Continuous performance measurement demonstrates the desire to achieve excellence in all aspects of business, from long-term planning to day-to-day process improvement. If you do not measure your performance, it is like playing softball or basketball without keeping score: you're only practicing. When it really counts, you keep score. For product development, you keep score by establishing metrics and monitoring your performance. In addition to keeping score, you can establish goals or targets for performance. Tracking performance against established targets results in a road map of your actual vs. desired capabilities, and provides a starting point for future improvement initiatives.

Mapping Quality Programs and Tools to the Strategic Quality Ideals

The Strategic Quality Ideals and Organizational Enablers have manifested themselves in a variety of quality programs and tools, including employee suggestion programs, high-performance work teams, benchmarking, customer action teams, vendor partnership programs, quality improvement teams, process management teams, Quality Policy Deployment, Quality Function Deployment, and more. Table 3.1 lists a few of the more well known programs, and shows their relationship to the Strategic Quality Ideals.

A common mistake is to confuse quality programs like those in Table 3-1 with the quality ideals. When this happens, failure of a quality program is associated with a bad quality principle. The quality movement is not just a "bunch of quality programs." It is a realization that certain fundamental ideals can have a serious impact on the long-term success of a business. The quality programs are tools for implementing these ideals in an organization. Dobyns and

TABLE 3.1 *Strategic Quality Ideals and Associated Quality Programs*

Strategic Quality Ideals	Example Programs or Tools
Clear Vision and Goals Communicated to All	• Policy Deployment—Florida Power and Light (FP&L) • Quality Policy Deployment—AT&T • Strategic Quality Planning—Baldrige Criteria • Constancy of Purpose—Deming's 1st of 14 points
Total Employee Involvement and Empowerment	• Employee Suggestion Programs • High Performance or Self-Directed Work Teams • Quality Control Circles—Ishikawa • Quality Improvement Teams
Customers as the Key Drivers for Change	• Customer Satisfaction Programs • Customer Action Teams—Milliken Company • Quality Function Deployment (QFD)—Clausing • Service Quality Indicators—Federal Express
Focus on Process: the How of Doing Work	• Quality in Daily Work (QIDW)—FP&L • Process Quality Management and Improvement—AT&T • Customer/Supplier Model • Statistical Process Control (SPC) • Benchmarking—Xerox • Q1 Supplier Certification Program—Ford • ISO 9000 Standards • Quality Improvement (QI) Story—FP&L • Plan-Do-Check-Act Cycle—Shewhart/Deming
Metrics for Continual Performance Assessment	• Six Sigma—Motorola • Zero Defects—Crosby

Crawford-Mason, in their book *Quality or Else*, write about Motorola, "What it didn't do is equally important. Motorola did not dash off willy-nilly, desperate to get something done. It didn't decide that quality circles or statistical process control or any of the other tools was *the* answer. In short, they didn't confuse technique with philosophy" (Dobyns and Crawford-Mason 1991, 128).

One pitfall you can avoid is to implement a program without introducing the fundamental principles behind the program. It is just as important to provide education on the ideals themselves as it is to provide training on any specific quality program.

ORGANIZATIONAL ENABLERS

Making the Strategic Quality Ideals work requires more than a belief in the ideals. Specific capabilities must be present in the organization, from leadership to appropriate reward systems. We have observed that the absence of these organizational capabilities can severely cripple the implementation of any quality program.

Leadership and Commitment at the Top

Strong leadership is essential if an organization is to successfully realize the Quality Ideals. The changes in culture and operation needed to do things the new way are too great to happen without strong leadership from the top. To avoid getting confused or overwhelmed by definitions of leadership and the attributes of a good leader, we list here the things a leader in quality needs to do:

- Drive the creation of a vision and long-term plan for the organization
- Articulate this view to all of the employees
- Communicate this message in such a way that people understand the directions, buy into the strategy, and want to follow
- Be a role model (walk the talk)
- Establish the values by which the company does business day-to-day
- Be willing to stay to course when things get rough
- Remove the barriers that keep people from being successful
- Keep the message ever present (manage change as a process, not an event)
- Rise above the day-to-day turmoil and keep hold of the vision
- Build excitement, loyalty, and motivation
- Walk ahead of the rest (people look for a leader)

Commitment at the Top is an important part of this organizational enabler. The programs that work toward the Quality Ideals take time and money to implement. They often involve changes to the organization, the culture, the value system, and individuals' roles—all the things that are most deeply rooted in the work environment. Real change (not quick fixes) for an organization does not happen as an event, but rather as a transformation that must have a strong, committed leader to drive the change.

Support and Alignment among Middle Managers

The degree of support and alignment among middle managers can have a dramatic impact on the level of success of many quality programs. The ability to achieve uniform success across the organization depends largely on the ability of middle managers to deploy programs that may not benefit them immediately, and in some cases to sponsor programs that go against their traditional view of managing the organization. Their roles may change, or their control may erode. We have observed cases where managers were actually working themselves out of a job.

The two Quality Ideals most impacted by middle management are employee involvement and empowerment, and focus on process. Employee empowerment is contradictory to some managers' view of the traditional role of employees vs. managers. To some managers, empowerment means giving up control, or being less involved in decision making. Some managers may see empowerment as giving up their "value-added" function in the business, and this surfaces the issue of job security.

Process focus has a similar effect, in that process management and deployment often usurp a traditional role of a manager: telling people how to do their job, and when. With process management, the *How* and *When* of the job are captured in process documentation, training courses, checklists, and other vehicles directly available to employees.

Poor alignment among managers regarding quality programs is not transparent to others in the organization, especially the employees themselves. In professional communities, this unrest is very visible, can be the topic in the halls or at lunch, and results in more time being spent on the controversy than on the programs themselves. People get caught up and find humor in the political tangle, and lose sight of the ideals that the organization is working toward.

In more extreme cases, managers may half-heartedly deploy concepts that they do not believe in. The net result of this action is that part of an organization goes through a transformation of roles and ways of doing work, and another part remains unchanged. When these two bodies attempt to work together, they find the differences in culture and values large barriers to success.

We observed one example between two large departments. One embraced the concepts of *Focus on Process: the How of Doing Work* and self-directed teams, while the other elected to maintain a more traditional approach to their work. When it came time for a new development project to begin involving both departments, the project team requested documents describing the processes. One organization could provide it, and the other could not. When the group attempted to work as a self-managed work team, they found themselves spending an inordinate amount of time working people issues between supervisors and project team members. One organization had gone through the transition necessary to support the new environment; the other had not. In many ways this created a dysfunctional team and complicated their work. The team members complained that they were spending their normal work time addressing process and people issues, and doing their project work in the evenings and on weekends. The overhead of the new environment was jeopardizing their project, and only through extraordinary effort and perseverance did this team rise above the turmoil to meet, and in some cases exceed, their objectives.

If middle managers in an organization suppress their political competitiveness, and together present a uniform view to the organization (even in areas where they personally may not agree), the result is a consistent, focused deployment of principles and programs, and there is a greater likelihood that employees will listen to the message and not focus on in-fighting within the organization.

High Morale and Satisfaction of Employees

It is hard enough to deploy fundamental changes in an organization when everyone is reasonably content with their work and environment. If employees are dissatisfied with their jobs or working conditions, and do not see the new programs improving these conditions, then the task of deploying quality programs becomes nearly impossible. The following example was relayed to us from an employee of a Colorado utility company.

Example. We [a group of utility employees] had been complaining about excessive overtime and safety issues for several months. We were brought into a meeting to announce the roll out of Quality Circles (an employee team approach to solving problems or introducing new ideas). Our response to management was, "we don't even have time to do our jobs let alone have a bunch of meetings to solve your problems, and by the way, you have not yet addressed our concerns about safety."

In this example, the employees came to the meeting with issues about their dissatisfaction with work, and management's new "exciting" program fell on deaf ears.

A worst-case scenario is when managers use the quality programs as a way to unload employee satisfaction issues onto the employees, and then do not support the resulting ideas. An employee for a unnamed company related the following story to us:

Example. We formed Quality Circles at work. We were not given any training, nor told what the purpose was, and the first idea we had was squashed by management. We had been struggling with overtime to complete our work over the past couple of months. One problem immediately surfaced by our group was the fact that the closest parking lot was a good 10–15 minute walk from the building. This walk not only ate into our time, but at night this was an issue for those of us (especially women) who were working late and had to walk that far to their cars. Our team recommended working on closer parking arrangements under certain conditions; that would have meant paying a local parking garage a fee for parking. Management refused to consider this option. What is the point of Quality Circles if our ideas are ignored, or they are not willing to put any money behind our improvement suggestions? No one really wants to do this [Quality Circles] anymore.

The net result was that employee satisfaction went down and the support for quality programs was lost, both at the same time. Imagine the result the next time the management of this company wants to roll out a new program.

This issue is relevant for the other quality ideals as well. For example, if employees are mistreated at work, and feel bad about their contribution and work relationships, then these employees will not be excited to hear about a program focused on Total Customer Satisfaction. They are preoccupied with their own problems, and if they have customer contact, they are more likely to transmit these ill feelings to customers than to work towards customer satisfaction.

This Organizational Enabler is the reason Federal Express upholds their "People-Service-Profit" philosophy, a belief that these three elements are interrelated and feed one another. The philosophy is that if you take care of your people (employees) and treat them well, they in turn will take care of and meet the needs of your customers. With satisfied customers, your business will be successful, and with this business success comes more job security and satisfaction for employees, and the cycle repeats. This interdependent triad acknowledges that employee satisfaction and morale play a key role in the ability of a business to delight customers and make money.

Strong Values Shared by All

Values are guiding principles that direct the behavior of everyone in the organization. They provide a common framework and assist in decision making. Values simply state the things that are valued by your organization. A good example is the Common Bond introduced in AT&T in the summer of 1992. The AT&T value statement includes:

- Respect for Individuals
- Dedication to Helping Customers
- Highest Standards of Integrity
- Innovation
- Teamwork

An example of the impact a value can have on a business decision is evident in the value *Highest Standards of Integrity.* This value implies that vendor kickbacks are not an acceptable business practice. It says that it is not acceptable to take bribes or payoffs from companies or governments in exchange for contracts. It implies that employees will not take credit for work completed by others. It demands that data that show the performance of a process or project are presented accurately and without bias. It states that everyone in the company will operate honestly and fairly in all aspects of their work.

Respect for Individuals also drives behavior in the workplace. Minorities, women, and the physically challenged must receive the same respect as other employees in the organization. Respect is given regardless of position in the company, whether you are talking about the president, a staff assistant, an engineer, or a mailroom employee. Following this value, many companies have adopted the term *associate* rather than *employee*. *Associate* communicates a new relationship between the company and the worker, and between managers and employees. Rather than saying "you are my employee," which carries with it the connotation that you work *for* me; the statement "you are my associate" says that you work *with* me.

Values such as these are essential for successful implementation of the Quality Ideals, specifically in the areas of employee involvement and empowerment. If everyone is to become a part of making the company a success, and dedicate themselves to quality in all aspects of their work, then each individual must know how to behave in relationship to other employees (or associates) and customers.

Reward Structure Linked to Contribution

When approached with a new idea or proposal for change, some people will be excited over the prospect of change and the corresponding opportunities. Others will not. For the most part, change is resisted and feared. Many people will assess the personal consequences of changing vs. staying the same. People make a personal decision to change, one at a time.

What motivates a person through a change process is a another book altogether. One observation we have made is that the reward and recognition system plays a role in this process. Failure to tie reward and recognition to contribution results in employees concluding that it may not matter whether they support the Quality Ideals or not. The converse of reward for contribution is the threat of punishment for lack of cooperation. Applying pain to gain the wanted changes is usually the last resort, and is less necessary with carefully managed change.

Employees and managers must know that the actions they take in support of the Quality Ideals are directly linked to their personal goals and success; they need to be able to identify with the opportunity. If they provide more support and act as role models, then the rewards and recognition for this behavior should be high (Chapter 14 is devoted to the topic of rewards and recognition). On the other hand, if they are nay-sayers and discourage the adoption of the Quality Ideals, then the rewards should be low. There should be no confusion among employees as to the tie of contribution to reward. If this is lacking, then telling people to focus on customers, be dedicated to quality, get involved to solve problems, etc., will most likely be met with superficial interest and long-term apathy.

In the most basic sense, you must ask if the behavior you want is being rewarded and reinforced. If the answer is no, then the old behavior will probably prevail. During the early stages of a quality transformation, the tie of compensation, rewards, and recognition to contribution must be high, as the behavior you desire early on is a greater change from the norm than you will observe years later, when the norm has changed and everyday behavior is based on the Quality Ideals.

The Strategic Quality Ideals and Organizational Enablers presented here set the stage for the rest of the book. At the beginning of each chapter you will find a list of the Ideals and Enablers addressed in that chapter. By the end of the book, you will have developed a sense for how the individual topics tie together to build a tool kit for achieving these Quality Ideals in your organization.

4 Improving Work Processes

Strategic Quality Ideals

✓ Customers as the Key Drivers for Change
✓ Focus on process; the How of Doing Work
✓ Metrics for Continual Performance Assessment

Organization Enablers

✓ Leadership and commitment at the top

The first three chapters set the stage for the basic quality principles. This chapter begins to apply these principles in the area of process management—a method to control and continuously improve *how* we do work.

Prior to having a process focus, our managers and experienced designers were the knowledge pool for how products were designed. First- and second-level managers led their groups through the development process based on their past experiences, both successes and failures. When a schedule was needed for a new project, one of these managers would create the new schedule by modifying an old schedule, often pulled from their personal files.

For many engineers, this schedule was the development process. It showed what would be completed and when. The less-experienced engineers on the project were forced to rely on a few lead engineers or managers to guide them through the myriad other development steps not included in the schedule.

Because this "process" knowledge resided with a few key managers, it easily eroded from the organization as these people transferred to take on other career opportunities or retired. New managers took their place and started the learning cycle over again. In many cases we repeated the same mistakes; all that had changed were the people.

This complete dependence on people to get the job done right has easy-to-recognize symptoms. First, when things go wrong, people, usually individuals, are blamed. Second, successes and lessons are seldom captured and shared across projects or between departments: mistakes repeat, and successes don't. Finally, no measures exist that allow project managers to assess the performance of their project against other projects, and in many cases they don't know if they are doing better or worse than their competitors.

When we operated in this mode, we had little idea if we were meeting the needs of our customers, inside or outside the company. We focused strictly on the product, and were insensitive to the methods and procedures we used. We did whatever it took to get the job done, and then moved on to the next project. Looking back, it is easy to see how internally focused we were.

A few forward-looking individuals in our organization were aware of these problems, and recognized a need to change. They hired an expert in process management to teach a course on the subject, and persuaded the top managers in the organization to attend a one-day class titled Process Quality Management and Improvement, or PQMI.

To the surprise of some of these new students on *process*, the case study for the class was a grocery store checkout line. The problem was with the line. People were waiting more than 10 minutes in the checkout line to pay for their groceries. "What does this have to do with product development?" we wondered. Most were humored by the situation, and since any group of engineers is challenged by a problem, on we went. We learned all about this small store's policies on cashing checks, marking prices on products, looking up codes for products without prices, and so on. Even hiring more checkers did not fully resolve the problems this poor store was having with long lines. Should they hire even more checkers? Should they change their policy for cashing checks? Should they make the customers bag their own groceries? We were supposed to help the checkout manager decide what to do.

The process management technique being taught in this course required that we write down the checkout process. Specifically, we had to identify the customers and suppliers along with their associated outputs and inputs, write down the process steps, and establish metrics based on customer needs. We were then provided with some data on the checkout process, and we went to work to determine the problem with the long lines.

As the day and class drew to a close, one of the two groups in the class was convinced they had the answer to the store's problem (that was our group). We were ready to go in and put everything right. We proudly presented our solution. To our surprise, we were wrong. The other group stated that their next course of action would be "to collect more data." Could that be right?

It was right. The instructor had purposely led us into the trap of not having adequate data to make a good decision. We, as faithful engineers, did the best we could with the data we had, and guessed. The other group did not. We went home feeling outsmarted, and maybe a little tricked.

It wasn't until some time later that we realized that the action we took in the course was exactly how we behaved in day-to-day work. We operated with insufficient data, both about customer needs and about our performance against competitors. We even solved problems and made decisions without all the facts. It was eye opening to realize that things could really be different if we could just make these *process* concepts work in development.

Seven years have passed, with both successes and failures to mark the journey. We have seen intervals reduced and quality go up, and gained a few gray hairs along the way. The competency of the organization is becoming balanced between excellent people and well-managed processes and tools. Today, we make more decisions based on data rather than intuition (although not all, we are still learning). We can't say that the grocery store example applied directly to a development environment, but it was a beginning, and what we have learned over this period we have tried to capture here. To give you a well-rounded picture, we first take a step back to look at some of the roots of process management.

BACKGROUND OF PROCESS MANAGEMENT

Every job is a *process.* In other words, you can break down any task into a series of steps that describe how to do that job. Moreover, you can measure and, most importantly, improve your work. Improvement can come from steady, incremental change, or it can come from breakthrough improvements. World-class companies have successfully applied process management techniques as a way to improve productivity and increase employee involvement in day-to-day work. Using process management, companies have

- involved employees in improving their own work processes,
- increased their focus on the customer as a key driver for change, and
- achieved breakthrough performance in time-to-market and quality of goods and services.

Process Management in Manufacturing and Service Industry

Many managers and engineers recognize the important role that process has played in manufacturing. Manufacturers around the world have embraced Statistical Process Control (SPC) as a common tool. In doing so, they have embraced the concept that a process can be characterized by a set of measurable parameters, and that *variation* of these parameters is a natural part of any process.

> *"Why are defectives and non-defectives produced together? The cause, as we stated before, is variation. Variation in* ***materials****, in* ***machinery conditions****, in* ***work methods*** *and in* ***inspections*** *are the causes of defectives.* (Kume 1989, 3)"

Managers and engineers have shifted their perspective on manufacturing processes. Instead of blaming people for quality defects, the *system* and the natural variation within the system are identified as the source of product defects. As a result, manufacturers have adopted new techniques for improving their processes as well as the way they manage the system—namely, the people and the processes they use.

While manufacturers strive to make products according to specific design specifications, service providers attempt to deliver services that meet the needs of their customers. Service providers work to understand customer requirements, and to meet or exceed customer expectations. They continuously change their service processes and offerings to adapt to the changing needs of their customers.

These lessons learned in manufacturing and the service industry can be applied to product development. The concepts of measurable parameters, or process metrics, and the system, or *process* perspective, provide two building blocks for a process improvement model for product development. The intense customer focus of the service industry adds another important building block. When taken together, these three building blocks from manufacturing and the service industry form the basis for improving work processes, or process management:

1. Any job or function can be viewed as a system or process.
2. Every process has an identifiable customer who has requirements and expectations.
3. Every process can be characterized by a set of measures, and variation in these process metrics is a natural part of any process.

Process Management for Product Development

Product development faces challenges different from those encountered in manufacturing or in the service industry. Some of these differences include:

- Longer process intervals in product development
- Greater variation from one development project to the next

- Different value system and culture inherent in product development organizations

First, manufacturing process intervals are typically short, on the order of hours or days, and these processes repeat many times over a short period. Manufacturing engineers and factory workers can collect data quickly, and tools such as control charts are very effective. For product development, process intervals range from weeks to years. The development of a new product, for example, ranges from six months to three years for most electronic firms.

Managing and improving processes in this environment requires a different mind set. It may take months to assess the full impact of a change in the front-end process or the design process on all downstream processes, whereas a change in the production process can be measured in hours. To complicate matters, you may have only a few data points to assess the effectiveness of process changes (only a few projects over a given period will use the development process), whereas in manufacturing you would be working with tens to hundreds of data points.

Second, in manufacturing, the same processes, with the same players, and the same inputs, repeat over and over again. Development projects change dramatically from one project to the next, with different people, requirements, and inputs each time. This variation creates new challenges for any process management model. These natural variations in the development process make it difficult to determine if the changes you made to the process resulted in an improvement, or if some other factor was the cause. In other words, it is very hard to change only one variable and note the effect. Moreover, the variation increases the further upstream in the development you go. The variation in the concept phase for a new product can be significantly larger than the variation in an assembly-line process in a factory.

Third, engineers and technical managers comprise the body of doers in product development. Unlike manufacturing, where managers can direct workers to perform their work according to a specified process, engineers in R&D tend to do what they want. In fact, many engineers are attracted to R&D because of an environment of freedom and creativity. Following a process is not immediately appealing to many engineers. Moreover, individuality and innovation are encouraged in an R&D community.

Developing process improvement models that address these fundamental issues for product development presents a real challenge. It is not simply a matter of saying, "here is SPC, go make it work."

PROCESS BASICS

A process model for improving productivity requires a perspective on work as a set of functions that can be combined to form a process. Contrast this to a perspective that focuses heavily on the *What*—the product itself.

Process is defined as one or more tasks that transform a set of inputs into a specified set of outputs (goods or services) for another person or process via a combination of people, procedures and tools. *Process Management* includes those activities surrounding the definition, deployment, measurement, and improvement of a process. The basic elements that characterize a process include:

- Process Purpose
- Customers and Suppliers
- Outputs and Inputs
- Process Boundaries (begins with and ends with)
- Process Steps
- Process Users—Doers
- Process Measures

The process purpose is a simple statement that summarizes what the process is designed to do. This purpose statement sets the stage for establishing the rest of the process elements.

Customers and suppliers of a process are those people or processes that receive output from, or provide input to the process, respectively. A process can be framed by its relationship to customers and suppliers, as shown in the Customer-Supplier model in Figure 4.1. This diagram is a simple way to illustrate a process, and helps answer the question Where does the process begin and end?

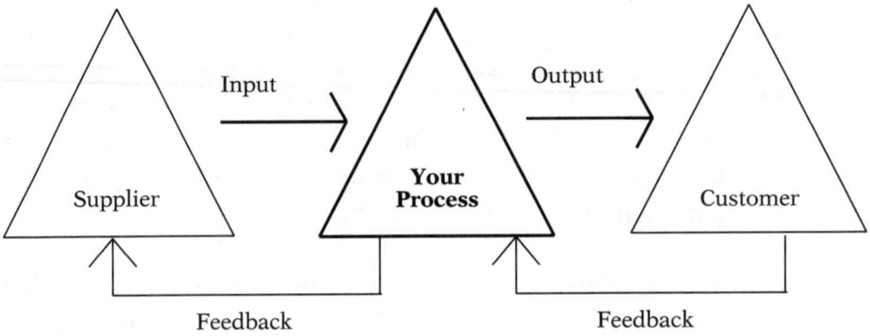

FIGURE 4.1 *Customer-Supplier Model*

Given this understanding of suppliers and customers, the inputs and outputs can be identified. Inputs are the deliverables received from suppliers. Outputs are the products of the process delivered to customers. The resultant set of information:

- Customers and Suppliers
- Outputs and Inputs
- Process Boundaries (begins with and ends with)

clearly defines the scope of the process. Once the boundaries are established, the set of work activities, or process steps, contained within these boundaries can be identified. Process steps are those activities performed between the process boundaries that transform the inputs into the specified outputs. Process users are those people who do the work described in the process—namely, those who transform the input into the output.

Process measures are the final ingredient that characterize the process. Process measures or metrics assess the performance of the process. They measure how well the outputs meet the customers' requirements, the efficiency or productivity of the process steps, and how well suppliers are meeting the needs of the process.

Process Basics: An Example

For example, take the development process used to review the design of an enclosure or chassis that will house an electronic circuit board, called here the Physical Design Review Process. The *purpose* of this process is to review the drawings for a new enclosure and to identify improvements in the design prior to building a prototype. The Customer-Supplier model for this process is shown in Figure 4.2.

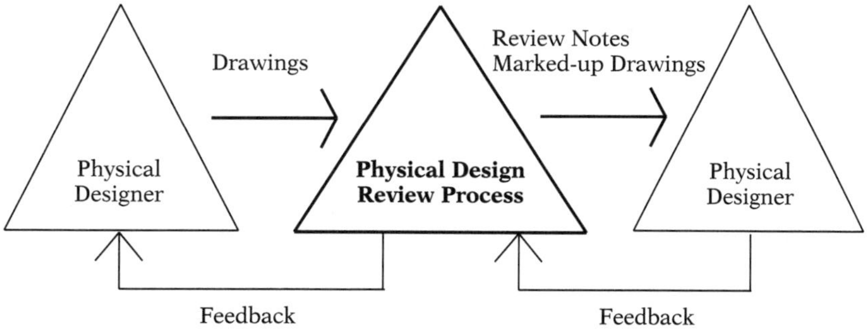

FIGURE 4.2 *Customer-Supplier Model*

The process *inputs* for this example are drawings of the enclosure. The process *outputs* are the notes from the review meeting, and marked-up drawings. The *process boundaries* can now be established. The process *begins with* availability of preliminary drawings for a new enclosure, and *ends with* improvement opportunities provided to the designer.

The process steps[1] for this process might include:

- Schedule a review meeting
- Distribute copies of the drawings prior to the meeting
- Conduct the design and manufacturability review
- Document improvement opportunities

The process measures for this process might include:

- Attendance level at meeting
- Number of improvement opportunities identified
- Number of improvement opportunities implemented
- Number of design faults found after prototype was built

Note that in this example, the customer and supplier are the same individual. Some process models say that you cannot be your own customer. We have observed that many sub-processes within development have this characteristic, and this should not be a concern for those applying these techniques in product development. We have found many people wasting time trying to develop a model for their processes in which the customer did not also appear as a supplier. Little effort should be spent agonizing over this dilemma. It simply is not important.

CONTINUOUS PROCESS IMPROVEMENT MODEL

Continuous process improvement as a mechanism for improving productivity requires that you have work processes that repeat with some frequency, and that the processes are similar with each subsequent execution. This requirement is certainly met in manufacturing, where the same work processes repeat frequently.

1. Process diagramming techniques are covered later in this chapter.

Continuous Process Improvement Steps
1. Document the current process
2. Identify customer needs and establish metrics
3. Set requirements on suppliers
4. Follow the process
5. Measure process performance
6. Identify opportunities and implement solutions
7. Repeat Steps 4–6

FIGURE 4.3 *Continuous Process Improvement Model*

Product developers also repeat the same processes. The added complexity for develoment lies in the content variation of the process inputs that occur from one project to the next. Fortunately, the other basic elements of most development processes remain the same. For example, the drawings reviewed in the previous example will vary according to the product size and required functionality (i.e., each time the drawings are of a different product), but the purpose of the process (to review and provide feedback), the basic review procedure (advance distribution and a review meeting), and the process outputs (meeting notes and marked up drawings) all remain the same. In many ways, the basic process elements you find in manufacturing processes are also present in product development. Figure 4.3 lists the steps we have found to be effective for continuously improving a process, whether it be a simple drawing review process or a complex product realization process.[2] In the sections that follow, these steps are described in detail.

Step 1. Document the Current Process

Process management is based on continuous improvement. That means that you start with what you have, and improve from there. In Chapter 5, we address process re-engineering—starting with a clean slate.[3]

To determine what you are starting with, you have to write down your current process. The process should be fully characterized (purpose, begins with, ends with, . . .) as described in the previous section.

2. Adapted from the AT&T manual, *Process Quality Management and Improvement Guidelines* (AT&T Quality Steering Committee 1988).
3. Chapter 5 also discusses which approach, process management or process reengineering, is more appropriate for your situation.

When we first chartered our design organization to write down their current processes, engineers had difficulty. The processes that were being written down were a combination of the way it was and the way they wanted it to be. To make matters worse, many could not agree on the current process. Engineers and managers had "done their own thing" for so long that there really wasn't just one process that they could agree on as the current process.

We countered this by having groups of individuals who worked in one functional area meet and write down what they did agree on and what they didn't. They would usually sort through their disagreements, and in most cases were able to come up with a process that was close to what was being done. In some cases people added items that may not have been done last time around, but that they intended to do next time. This compromise was the only alternative given that the documented process needed to reflect what they really intended to do.

The next dilemma we encountered was how to write the process down. In manufacturing, many processes were written down as text procedures (no pictures). The process tools we were using, however, offered two other alternatives: flowcharts and block diagrams. Some people liked flowcharts; others liked block diagrams. At first we encouraged engineers to use just one method. Later we learned to be more flexible, and discovered that the best tool for the job depended on the process they were documenting.

For example, consider a high-level process used to take a product idea and turn it into a product (often referred to as a *Product Realization Process*), and a low-level functional process for reviewing drawings. Figures 4.4 and 4.5 show simplified versions of these processes using two methods of process diagramming: a simple flowchart and a block diagram.

The method you choose should depend on the level of the process you are describing and the customer (the user) of the process documentation. Flowcharts are usually better for high-level processes that span a wide range of activities. Flowcharts show temporal relationships well, easily illustrate high-level steps, and are fairly simple to understand. Block diagrams are useful when you need to show the organizational relationships for the process (i.e., who does what when). They take a little longer for most people to understand than a flowchart.

The process diagrams in Figures 4.4 and 4.5 are not the only ways to document a process. You may elect to document the entire process with text. How the process is captured is less important than the accuracy and clearness of the

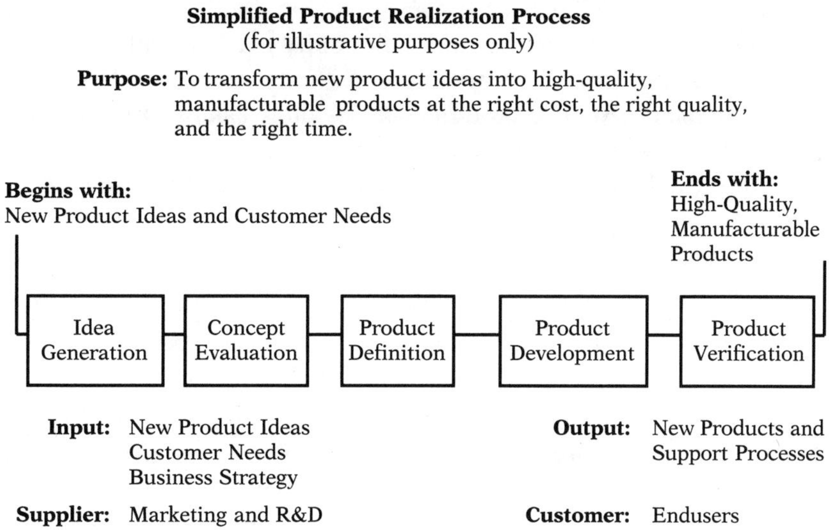

FIGURE 4.4 *Simplified PRP Process: Flowchart Example*

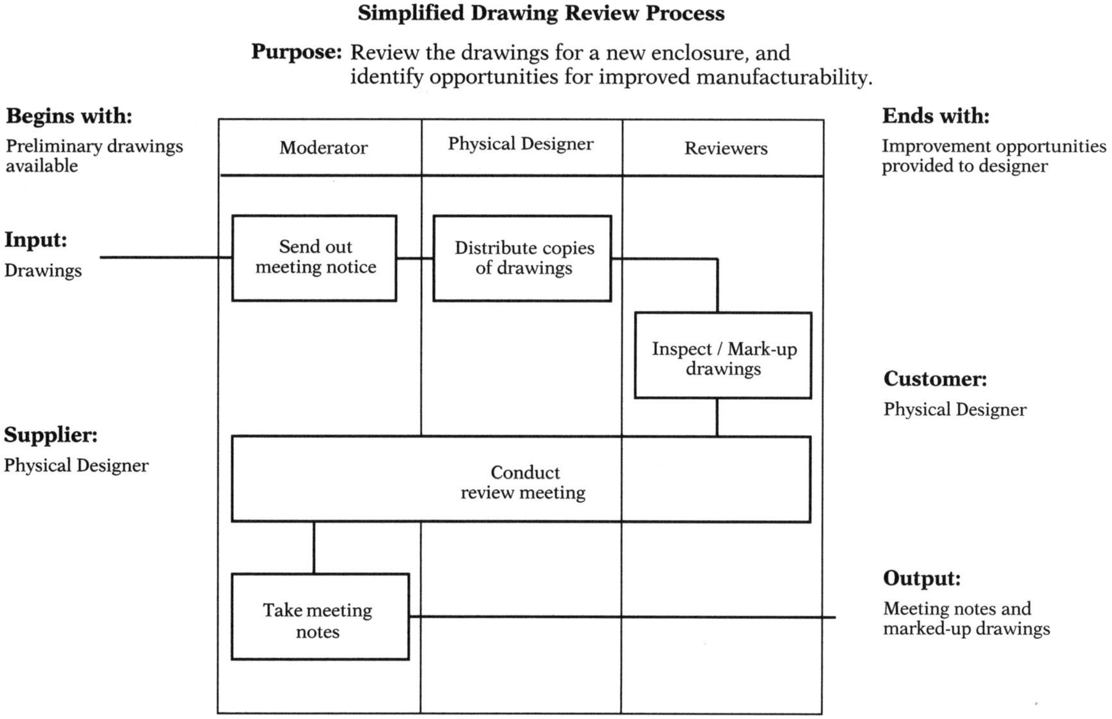

FIGURE 4.5 *Simplified Drawing Review Process: Block Diagram Example*

description. In Step 4 of this improvement model you will ask people to follow the process consistently, and their ability to do so will depend heavily on your choice of process documentation methods. Ask yourself: Is this process easy to understand and use? You may also want to ask the process users (remember—they are your customers!) if your process documentation is easy to understand and use. Three important considerations in documenting any process include:

1. Clarity of process description
2. Accuracy of process elements (inputs, outputs, steps, . . .)
3. Ease of use of documentation, from the perspective of the process users

We have observed process documentation ranging from one page to an entire 2-inch-thick notebook. Our experience with long, detailed process documentation has been poor. People just don't pay attention to it. If they cannot read and understand it quickly, they won't follow the process. We recommend that you keep your documentation down to the minimum that conveys the information.

Process Checklists and Gates

We have found checklists and process gates to be effective process documentation. A checklist is simply a list of items that are completed as a person follows the process. A process gate is a point in the process at which certain criteria must be met before the user can continue. Long difficult processes can be made much easier for the user by employing several process gates (also called *quality gates*) that have "passing" criteria identified for each gate.

In our current development process we have five process gates that occur between the idea phase and the production phase of a project. At the idea phase we may have five to ten high-level requirements for passing a gate, whereas at the release-to-manufacture gate we currently have more than a hundred items. We have found these types of gates to be very helpful for ensuring that requirements for moving on at each phase are known and met, and that the business justification for the project still exists. More examples of process gates and gate criteria are provided later in this chapter.

Process Hierarchy

In Figures 4.4 and 4.5, we showed one high-level process and one low-level process. Implicit in describing these processes as *high level* or *low level* is the principle of process hierarchy, or multiple levels of process description. Imagine documenting a product realization process from idea generation to product introduction at the detailed level shown in Figure 4.5 for the drawing review process. The result would be a very long description that combined very detailed functional activities with very high-level project management activities. Such a process description would be awkward and hard to follow.

You can be much more effective—and faster—in documenting your processes by employing multiple process levels. For example, note that in Figure 4.4 the entire product realization process is pictured with only five major steps. Inherent in this type of high-level description is that another process description exists for each of the boxes. These more detailed processes describing the activities in the boxes are called *sub-processes.*

As an example, for the box labeled *product development* in Figure 4.4, you could document the relationship between major development processes such as software development, hardware development, plastics development, etc. in your first sub-process diagram. Then you can use another level of sub-process to show the actual development activities in each of these areas. For this example, you would end up with three levels of process descriptions:

- Level 1 = Product Realization Process (like that shown in Figure 4.4)
- Level 2 = Major Product Development Processes
- Level 3 = Detailed functional processes (at the level of detail shown in Figure 4.5)

We have found benefits to employing a process hierarchy other than the ease and simplicity of writing the process down. First, we were able to document the processes for specific customer groups (this is important for a step described later in this chapter). For example, the process for software development could be separated from the other functional processes, and tailored for software developers. Second, we found that different people are needed to describe the process at the different levels, and identifying the right people is easier once the process is partitioned and leveled in a reasonable way.

Roles and Responsibilities of Process Users

Who does what when? The roles and responsibilities of the process users should be documented along with the process steps. For detailed functional processes, this may be as simple as noting the role along with the process step. For example, in a software design process one of the process steps is to "write code." The role here would be "software designer." It may not be necessary to write any more than that. Another process step in the software design process may be a "code review." The roles here may include the software designer and peers of the software designer with expertise in the appropriate area for the review. In this case, you may want to document the expectations (before and during the review) of the reviewers.

For higher-level processes, such as the overall product development (or product realization) process, the description of the roles and responsibilities may be more complex. For our product development process, we have a team management model. This team management model includes descriptions of the following roles and responsibilities:

- The project management team (the core team)
- The full team of project members (the extended team)
- The functional managers (supervisors of the core and extended team members)
- The senior managers (leadership for the business)

The descriptions of these roles and responsibilities required as much work as writing down the process steps, and about as much paper. We invested the energy to do this because we found that the need for clear roles (for people and teams) in the process was as important as the process description, especially when it comes to following the process.

Step 2. Identify Customer Needs and Establish Metrics

At the heart of good process management is focus on the customer. This focus shows up in the form of metrics that assess your ability to meet your customers' needs. Developing effective metrics that are specific, measurable, actionable, relevant, and timely will require some effort on your part, as well as some background on metrics in general.

Process measures generally fall into two categories:

- Performance measures
- Diagnostic measures

Performance measures are high-level metrics that assess the overall performance of the process. They are external in nature. Performance measures are most closely tied to outputs and customer requirements, and to business needs for the process. Time-to-market, customer-found faults, cost of goods (material, labor and load for the product), and customer satisfaction are examples of performance measures.

Diagnostic measures are measures that ascertain *why* a process is not performing up to expectations. They are internally focused. Diagnostic measures are usually associated with the process steps and inputs received from suppliers. For time-to-market, diagnostic metrics might include the interval for each functional sub-process in development, or the interval for each major function in the product realization process (concept evaluation, product definition, product development, etc.). In the area of product quality, diagnostic measures might include system test results, component reliability ratings, and MTBF (mean time between failure) analysis for the product.

Catch-22 for Product Development

A common mistake with metrics is having too many. On the other hand, if you don't have enough, and the process performs poorly, you may not know what to fix. You may want to begin with performance metrics, and only collect a minimum set of diagnostic measures. If the process performs well and meets your customers' and business needs, then diagnostic measures become less important. If the process does not perform well, then a comprehensive set of diagnostic measures can be temporarily put in place to help identify opportunities for improvement.

For example, consider the simplified product realization process shown in Figure 4.4. The performance metrics might cover three general areas: time-to-market, cost to develop, and customer satisfaction with the product. If objectives for each of these three metrics are achieved, then further diagnostic measures may not be needed. On the other hand, say the cost and time-to-market objectives were not met. Each sub-process may now be evaluated to identify outages and surface improvement opportunities by looking at in-process, diagnostic measures.

The catch-22 here is that product realization intervals are typically long (i.e., measured in months or years). Hence, if diagnostic measures were not collected at some level during the process, then insufficient data may be available to help you understand the problems. You probably do not want to wait until the entire cycle repeats before you collect more data and institute improvements; herein lies the catch, and also a major difference in manufacturing vs. development processes.

The balance that you need to strike is to establish enough diagnostic-level measures to point to the opportunity, much like a triage evaluation. You may not know exactly what went wrong, but you will have framed the problem and can begin working on the most critical areas first.

Establishing Metrics

To establish a set of metrics for a process, a simple Five-Step method is suggested in Figure 4.6. Because customer requirements drive performance metrics, the first steps are to understand the needs and requirements of your customers. You should actively engage customers to determine their needs and expectations, and to understand their requirements. Business needs will also drive performance metrics. For product development, these business needs should be captured in a goal-setting process, like that described in Chapter 9.

Some caution needs to be applied when metrics are based solely on customers' "voiced" needs and requirements. Step B of the process in Figure 4.6 suggests that customers be interviewed or surveyed to determine their needs. However, customers may not voice "expected" attributes of a product or service. Customers take these items for granted, or consider them commonplace. Customers also will not voice product or service attributes that they are not aware of—things that surprise and exceed their expectation. The observation that many customer needs are not verbalized is an important consideration as you determine your customers' requirements.

Establishing Process Metrics

Step A. Identify customers and outputs of the process. Customers may include:

- End-Users of products and/or services
- Process managers of downstream processes
- Process users

Tools: process block diagrams or flowcharts (from Step 1)

Step B. Determine customer needs / requirements

- Review outputs with customers—gain buy-in
- Determine their needs and requirements
- Ask *them* how to measure how well you are meeting their needs

Tools: Interviews, Surveys, Role Reversal

Step C. Understand key goals for product development

Tools: goal-setting process (see Chapter 9)

Step D. Determine effective measures, including both performance and diagnostic metrics.

Tools: brainstorming, Affinity Diagrams (see Chapter 6)

Step E. Compare / filter / align metrics with higher-level processes (especially true for sub-processes)

FIGURE 4.6 *Method for Developing Process Metrics*

Kano, Seraku, Takahashi, and Tsuji characterized three major elements of quality in their article "Attractive Quality and Must-Be Quality" (Kano et al 1984, 4). They classified these primary elements as:

- Must-be quality—results in dissatisfaction when not fulfilled, but does not necessarily produce positive satisfaction when fulfilled (product safety would fall into this category)
- One-dimensional quality—results in satisfaction when fulfilled and dissatisfaction when not fulfilled (product life for a television would be a one-dimensional quality element)
- Attractive quality—produces satisfaction when fulfilled, but is acceptable as is when not fulfilled (new technology and cosmetic appearance would fall into this category)

One way to address the complete set of customer needs is through processes of role reversal—walk a mile in their shoes (and they in yours). The underlying assumption here is that if you understood your customer's job or function as well as your own, you would be in a unique position to deliver attractive quality, while at the same time satisfy the one-dimensional and must-be quality elements. Some programs that support this type of role reversal include job rotations (e.g., a design engineer takes a temporary assignment in manufacturing, services, or sales support, and vice versa), and cross-functional training (engineers study the processes and functions of their customers). The goal should be for every supplier to have enough knowledge of their customers' processes and tasks that they develop a profound understanding of their customers' needs.

As you work through the process for establishing metrics, care should be taken to check against the following criteria:

- **Drives right behavior**. People will take action to achieve the objective for the metric. In some cases, the action they take may surprise you. The question to ask yourself is "Will this metric drive the desired behavior?" or "What type of behavior might this metric drive?"

Example. An example we experienced involved a metric to assess the number of customer-affecting design changes, known as *Class A* changes. The purpose of classifying a change as Class A was to ensure that the right processes were followed for changes that had an impact on product in the field. The other change class that was used was *Class*

M, changes that only affect product currently in manufacturing; these changes have no impact on product in the field.

Because of the impact on product in the field, Class A changes cost money and were a poor reflection on the design and manufacturing communities. Designers and manufacturing engineers would go to great lengths to avoid a change being considered Class A. Meetings would focus on whether a change could be classified as M rather than A. An environment of fear was created—and in the wrong place. Engineers were afraid to issue Class A changes because it made everyone look bad, including them.

The Class A metric was intended to evaluate the quality of the initial design (i.e., the processes that were being measured were development processes). Unfortunately, the processes and people that were affected most by the metric were those doing current engineering (maintaining the product through its lifecycle). Some of these people were not even involved in the original design. Their goal should have been to do whatever was required to ensure we met customers' needs. Instead, they attempted to minimize the number of Class A changes to optimize the *metric*. The Class A metric drove behavior away from focusing on the customers, and focused instead on "what can we do to avoid considering this a Class A change."

Many engineers felt that metrics like these added no value, but at the same time felt obliged to work within the system. Other more outspoken engineers complained, and the metric was eventually dropped. It made no sense to have a metric to assess design quality when it drove behavior in a direction opposite the best interest of the customer.

- **Easy to understand.** Metrics should be simple. If they require a lot of explanation and definition, then collecting data and translating the data into actions becomes more difficult. Easy-to-understand metrics are easier to sell, and have a stronger impact on the process and the process users.
- **Specific.** Metrics should be specific to the need they're measuring. Broad metrics that cover too many customer needs are not as valuable as specific metrics that can be closely associated with a single customer need.
- **Measurable.** It is possible to develop metrics that cannot be measured, or that cannot be measured without extreme cost or effort. Consider how the data will be collected, and if the data collection process is feasible.

- **Accurate.** Try to avoid measures that are not accurate or not controllable. Having bad data for a metric is worse than not having the metric at all. If accurate data from a controllable source are not possible, consider a different metric.
- **Timely.** Metrics should be collected in a timely manner. Any data on a process that come months or years after the process is complete may be ineffective at driving change and assessing the process that is in use today.

Step E in Figure 4.6 suggests that the metrics for the process should be aligned with higher-level processes in product development. This step is especially important for sub-processes that may unknowingly be sub-optimized, thus compromising the performance of other processes.[4] We encountered this problem during the first several years of applying process management techniques in product development. Each of our functional organizations initially went off to work their own process. They developed measures and optimized their own processes independent of the impact on the overall product realization process. We countered this by applying policy deployment methods (covered in Chapter 9), and by deploying the metrics and objectives of the product realization process to the functional sub-processes. Figure 4.7 shows the customer-supplier model in a series of processes. Each triangle in this figure represents a process. At the same time, each triangle in this series is first a customer of the previous process, and then a supplier to the next process. Figure 4.8 shows this same set of processes as a subset of a higher-level process.

If the only process structure in place is like the structure shown in Figure 4.7, then sub-process optimization is likely to occur. If, on the other hand, key metrics and goals are also deployed from a higher-level process as shown in Figure 4.8, then alignment can occur both between customers and suppliers (horizontally), and between processes without a direct interface (i.e., processes

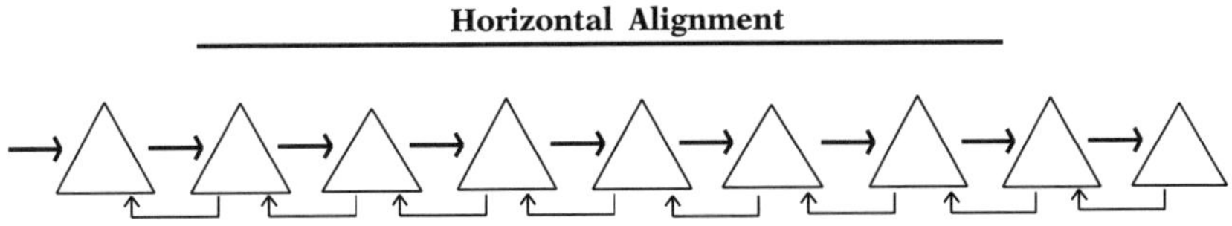

FIGURE 4.7 *Horizontally Aligned Process Structure*

4. An example of a sub-process would be a plastic design process that is a sub-process for an overall equipment development process. At the same time, the equipment development process is a sub-process for the product realization process of an electronic product.

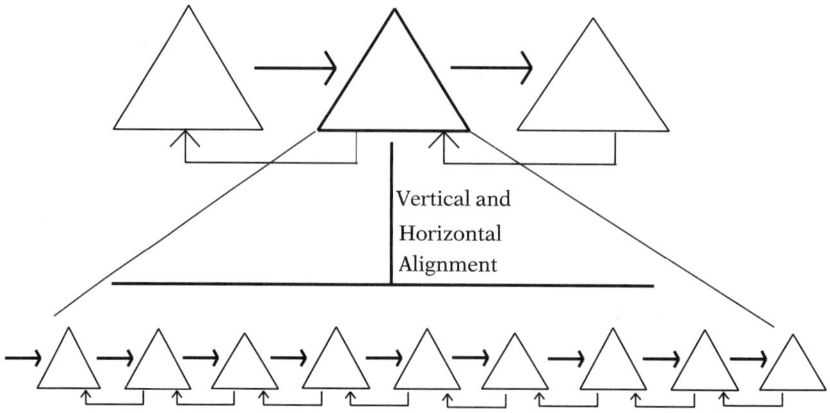

FIGURE 4.8 *Horizontally and Vertically Aligned Process Structure*

that do not have a direct customer-supplier relationship) via a vertical alignment with the higher-level process. Said another way, you can look *downstream* for the needs of your customers, and look *up* for the needs of higher-level processes, as well as for the needs of the business as communicated through goals for product development.

When you have completed identifying metrics, you should step back and do a sanity check. Questions you can ask include:

- Do these metrics make sense?
- How do these metrics compare with any existing metrics?
- Is it a complete set (e.g., have we adequately covered the areas of time, quality, cost, and customer satisfaction)?

The metric areas of time, quality, cost, and customer satisfaction seem to be generic across industries. Consider the key objectives of Lands' End, a well-known direct-order clothing company. Their key objectives from the Spring 1993 catalog read:

1. Make your merchandise as good as you can.
2. Always, always price it fairly.
3. Make it a snap to shop for, 24 hours a day.
4. Guarantee it. Period.

You can clearly see that the elements of time (Item 3), quality (Item 1), cost (Item 2), and customer satisfaction (Item 4) are evident in their objectives as well. Product development concerns that map to time, quality, and cost include time-to-market, product reliability, and cost of goods. In the area of customer satisfaction, product development concerns itself with end-users' satisfaction, and with downstream customers such as manufacturing and services.

Once you have established process metrics, objectives can be set. Objectives should be based on your customers' expectations and on benchmarks from competitors. These objectives may or may not be based on your present performance. Depending on the amount of historical process data that you can collect, you may not have sufficient data to show your current performance. That should not stop you from establishing objectives that make your process best-in-class in meeting the needs of your customers. Be careful not to get bogged down in what data you already have and what you don't. Continually focus on your customers and competitors, and base your objectives on them.

As you begin to follow your process consistently (coming up in Step 4 of Figure 4.3), you will begin to develop current performance data that you can compare with your benchmarks and objectives, and from this comparison you will identify areas for improvement.

One more note regarding metrics: check to be sure that your process metrics and objectives measure results, and not the activities leading up to results. Many process improvement efforts fail because they are unable to demonstrate results. They instead focus on their rate of activity, number of processes documented, or number of people trained. Focus your metrics on the results you and your customers expect to see.

Step 3. Set Requirements on Suppliers

In the same way that your customers' needs and requirements were important for establishing metrics and improving *your* process, requirements on your suppliers are equally important—both for them and for you. For them, these will be the requirements of *their* customers, and if they are applying similar process improvement strategies, then they, as suppliers, need this information. For you, setting requirements on suppliers ensures that the inputs to your process meet your needs, thus enabling you to transform the inputs into the desired outputs for your customers. This avoids the scenario of "garbage in, garbage out."

A strong link should be evident between the requirements placed on you by your customers, and the requirements you place on suppliers. The metrics for how well these suppliers meet your requirements should be generated by, collected by, and analyzed by your suppliers, not by you. It is well worth your time to help your suppliers understand the importance of measuring the quality and acceptability of their output. This focus places accountability and ownership with them, since it is their responsibility to change when their inputs are not meeting the needs of your process. You should meet with your suppliers, and together agree on and document supplier requirements.

As a good customer, you have an additional obligation to provide feedback to suppliers. This feedback includes providing information on the quality of their inputs from actual experience in your process, and returning inputs (products or information) that do not meet your requirements. Supplier relationships are a very important part of the process improvement model, and the better your partnership with your supplier, the better the input (materials and services) will be to your process.

Step 4. Follow the process

Sometimes the most obvious step in any process is the most elusive. In the case of process management, following the process is such a step. At this step, the person managing the process must rely on the actions of others to continue the process improvement cycle. Many people working on process management become frustrated at this stage, especially when process users (the doers) ignore their process. Many engineers prefer to develop their own process as they go, and may not be receptive to following what they believe to be prescriptive steps to accomplish their creative work.

In response, a process manager may attempt to police the process users to ensure that the process is followed (we tried this). Development engineers find this approach very distasteful. Other process managers may put forth their process as a guideline, allowing each process user to tailor the process to their specific needs. This approach definitely produces less friction, but if each execution of the process varies too much, subsequent process improvement is difficult. It is hard to improve a process when you don't know what process your associates are following, and if the process is different each time, then the process improvement model falls apart.

Several methods we have found effective for ensuring that process users follow the process, yet remain sensitive to the need for a free and creative environment in product development are:

- To communicate clearly the benefits the process brings, both to the organization and to the users

- To provide effective training and easy-to-use process documentation
- To use exit/entry criteria or the concept of gates in the process
- To gain the support of management, and ensure that this support is communicated widely

Process managers must become salespeople to some degree. What they are selling is their process and the associated benefits of using it. They should work toward a process that everyone *wants* to use. If process managers cannot clearly communicate the benefits, then many engineers may be unwilling to use the process. The more data available to substantiate these benefits, the better. If you have success stories or testimonials, use them. If you have evidence from other companies using similar processes, use that. Make it clear to the users that the value in following your process can be quantified in reduced costs, improved quality, reduced time, or improved work life for them. They will be much more willing to change once they buy into the benefits.

Providing large notebooks of process documentation to your development community can be a turn-off. If the process is hard to follow simply because of the documentation, then people will resist using the process. Offer effective training and easy-to-use process documentation, and they will respond. The easier the process is to follow, the more people will follow it. Process training should begin with the benefits discussed previously, and then move to the conceptual elements of the process. Conceptual elements include basic principles that the process follows. For example, the drawing review process discussed previously was designed around two fundamental concepts: advance distribution of the drawings for review, and a formal review meeting to discuss issues. Process users will find it helpful to understand these conceptual elements before diving into detailed process diagrams. Follow up the conceptual presentation with a detailed presentation of the process, including all of the process basics discussed at the beginning of this chapter.

Entry and exit criteria help people follow the process. Entry and exit criteria are essentially checklists, preferably on one sheet of paper, that help process users understand if the required inputs are available at the beginning of the process, and if the required steps, with associated outputs, were completed during the process. You might think of them as reminder sheets. Exit criteria are especially useful in reminding users to complete steps that they might have overlooked, and to ensure that the customer's requirements have been met.

In-process criteria are often referred to as *process* or *quality gates*, as discussed previously. Process gates help people verify that the necessary activities were completed, and necessary elements are present, before moving on to the next

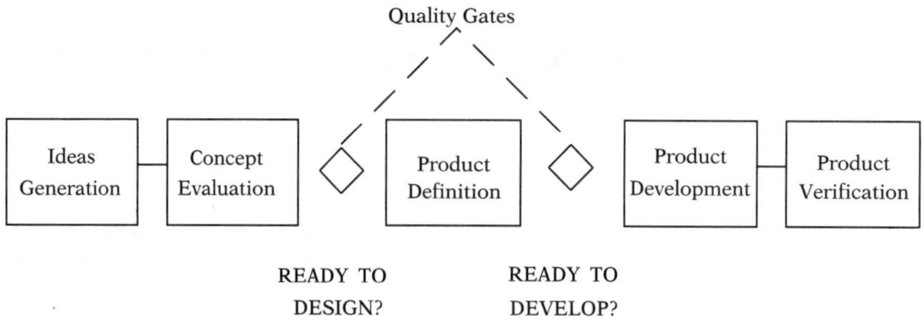

FIGURE 4.9 *Example of a Two-Gate Product Realization Process*

phase in a process. A simplified example is shown in Figure 4.9 for a product realization process.

The diamonds in Figure 4.9 represent the process gates in the process. At these gates, the quality and completeness of the work up to that point is evaluated, and a go or no-go decision is made. At the Ready-to-Design gate, for instance, the criteria for passing through the gate might include:

- Viable business case and marketing plan
- Complete list of customer needs associated with the product concept
- Volume and cost targets for the product
- Key product requirements and functionality listed in priority order
- Product and manufacturing technology readiness

You are certainly not limited to only two gates. Gates may be added anywhere in the process as needed. Good places for gates in the process include wherever your product realization process spans organizations, wherever major resource or investment decisions are made, or wherever major hand-offs occur. We have found these checkpoints, or quality gates, to be useful when deploying our processes. They are also effective as markers to measure time-to-market.

Quality gates will also enable you to involve senior managers in the product realization process. It will become evident throughout this book that we are encouraging more and more of the daily decisions to be made at the lowest possible level. Process gates provide an opportunity for senior managers to be involved in critical decisions for the business, without being tied up with the day-to-day activities of managing a new product.

Clearly stated benefits, effective training and documentation, and process techniques such as quality gates are all important; however, none of these will be sufficient unless they are backed by management support. You should make every possible effort to solicit support, both for specific processes and for endorsement of a process improvement system. All of us are resistant to change. The authors have been amazed at the power of a few simple words from management to spur people to cooperate and participate in process-focused activities. Without management support, the questions and comments people make about process include:

- Why do it?
- I don't have time to be bothered with this.

With management support, the questions and statements you hear include:

- How can I learn about the process?
- Where can I go for support for my process work?

In industry today the annual or periodic performance review remains a primary motivation for many individuals, and this is equally true in product development. Direct supervisors, who conduct these reviews, play a key role in setting priorities for engineers. The resulting influence of these managers on the success or failure of any process improvement program cannot be understated.

ISO 9001 registration for product development can also be helpful with process compliance. ISO registration is becoming a ticket for entering markets around the world. People readily identify with these new markets as representing more revenue, and therefore they buy in more quickly. ISO registration requires that your processes be documented, that you follow those processes, and that you are able to produce objective evidence in the form of quality records and documents that demonstrate compliance. ISO registration also requires that you have a functioning corrective action loop, much like that described in this continuous improvement model. ISO can serve as a strong leverage for process management because it introduces a more concrete driver, one that people identify with quickly. We have found that ISO compliance and process management go hand in hand and help an organization succeed with general process deployment activities.

Step 5. Measure Process Performance

If you are successful at Step 4, then two avenues are open to you to assess the performance of the process:

- Conduct project retrospectives
- Collect and analyze process metrics

A project retrospective is a review and evaluation of a project. Process users (the doers) gather with process managers to discuss their successes, their frustrations, breakdowns in the process, and whatever else is on their minds. Project retrospectives are usually a one-time event at the end of the project. However, for high-level processes that have long intervals, additional retrospectives may be conducted at key points during the process. These intermediate retrospectives are conducted to avoid loss of learning that may occur if a long time elapses before people are asked about specific parts of the process. These intermediate reviews may align with the quality gates discussed previously. For detailed sub-processes, the retrospective would usually occur after that specific process was completed.

Retrospectives typically involve a meeting. People working within a project come together with people interested in improving the process, and they discuss issues, lessons learned, etc. The moderators for the retrospective may request that all of the involved parties prepare a written list of issues to discuss prior to the meeting in order to make the process more effective. During the meeting, several techniques may be employed to bring as much information about the process to the surface as possible, including brainstorming, root cause analysis, and Affinity Diagrams. The moderator should spend equal time between project mistakes (i.e., if I had it to do over again, I would do this differently) and project successes (i.e., this really worked for us).

Unlike retrospectives, the collection of data on process metrics should occur during the process, not after. It is very difficult to look back at a project and compile accurate, reliable data. Process managers should work with process users to make the data collection easy and, when possible, automatic. For example, if some of the process metrics involve intervals or times between activities, then the process manager can work with the process users to ensure that their normal project schedules include milestones at the right places, so the interval data are collected automatically as the project manager tracks the progress of the project.

Another way to make data collection easy is to include data collection with the exit/entry criteria discussed previously. These checklists can include questions or templates for collecting metric information.

The results of the data for a project should be evaluated against the established objectives for the metrics. Areas with process performance that fall short of the objectives are prime candidates for more study and subsequent process improvement (Step 6).

These two methods—project retrospectives and data collection—are not completely independent. It a project has met or exceeded the objectives established for the process metrics, then the focus of a retrospective will be on gathering success factors from this project. If, on the other hand, performance in a specific area was poor, then a project retrospective could focus on lessons learned from working through problems on the project. In most cases, project retrospectives will be a combination of capturing success factors and learning from project mistakes.

Step 6. Identify Opportunities and Implement Solutions

Process improvement happens here, at Step 6. Given the data on process performance and the input from project retrospectives, the process managers are responsible for changing the process to improve performance.

In the case of success factors surfaced in project retrospectives, the process managers may be able to incorporate the learning into the process simply and directly, especially if the process users were able to communicate fully the details of their success.

In the case of project problems, or poor performance as indicated by process metrics, identifying process improvements will require a systematic approach to determine the root causes and develop solutions. We describe such an approach in Chapter 6. In special cases, the process performance may be bad enough to warrant a complete re-engineering of the process. We address process re-engineering in Chapter 5.

Project retrospectives and performance data are not the only stimuli for process change. Periodic benchmarking is another important tool for improving work processes. Benchmarking is addressed in Chapters 5 and 9.

Process improvement opportunities in a steady state mode are therefore derived from three primary areas:

- Process performance data (from the metrics you selected)
- Project retrospectives (actual experience about what works and what doesn't)

- Benchmarking (keeping an eye on what others are doing to improve their processes)

Using these three sources, you ensure responsiveness to customers, process users, and your competitors. Process performance data shows you how well you're meeting the needs of customers. Project retrospectives allow process users to share their experiences, and benchmarking keeps you in touch with your competitors or others who perform similar work processes.

Step 7. Repeat Steps 4–6

The process management steps described here, namely Steps 4–6, repeat over and over again. The underlying concepts become a part of the culture and the management system, and over time the day-to-day work and the process management work are indistinguishable.

How will you know when you have achieved success at process management? You will have easy-to-follow documentation for your development processes. Engineers will follow these procedures to develop new products. Continuous and breakthrough improvements will be evident from ongoing process improvement activities and process performance metrics. And, most importantly, your customers will be satisfied (sometimes delighted) with the goods and services you provide.

5 Process Re-Engineering

Strategic Quality Ideals

✓ Customers as the Key Drivers for Change
✓ Focus on Process: the How of Doing Work

Organizational Enablers

✓ Leadership and Commitment at the Top

INTRODUCTION

Like process management as described in Chapter 4, process re-engineering focuses on process as a way to improve day-to-day work functions. Process re-engineering, however, approaches process improvement differently, and on a much shorter time scale. While process management is based on the concept of continuous, evolutionary improvement, process re-engineering searches for breakthrough improvements and revolutionizes the methods for doing work. Re-engineering efforts tear apart the current process and systematically rebuild it. For product development, this may mean redesigning something as small as the design review process, or re-engineering something as large as the end-to-end product development process.

Many engineers with whom we have worked prefer process re-engineering to process management. Process re-engineering happens faster, creates something new and different, and gives people a real sense of accomplishment in a very short time. Creative designers or engineers feel at home with process re-engineering, and find it challenging and rewarding.

People often ask "When would I do process re-engineering vs. process management?" This depends on your current situation. If your process is performing well, with no real outages, then process management may be appropriate. If the process is performing poorly based on feedback from customers, comparative data with competitive processes, or complaints from users, then process re-engineering may be needed.

Another factor influencing this decision is whether a process management structure is already in place. You may find it easier to repair an existing

Process Management		Process Performance: Poor	Process Performance: Acceptable and improving
	Effectively implemented	Use problem-solving tools in most cases (Chapter 6)	Continue with process management (Chapter 4)
	Not present	Implement re-engineering (Chapter 5)	Implement process management (Chapter 4)

FIGURE 5.1 *Process Health Matrix*

documented process through normal process management techniques (i.e., problem solving) than to re-engineer the process from scratch. Only when the process is severely broken would you consider re-engineering a process currently under a process management structure.

Figure 5.1 shows a decision matrix that attempts to capture this thinking graphically. The y-axis reflects the state of process management, and the x-axis reflects the relative performance of the process. If no process management is in place and the process performs poorly, then you would select process re-engineering (lower left quadrant). If process management is not in place, but the process performance is acceptable, then select process management (lower right quadrant). If the process is not performing well but is under a process management and improvement structure, problem solving may be the appropriate course of action (upper left quadrant). In this case, it is important to evaluate the seriousness of the performance problem. Major process performance problems may be more effectively addressed with process re-engineering.

An exercise often used by our teams during team-building classes helps show the difference between process management and process re-engineering. A group, usually about ten to forty people, gather in a circle. They are given a tennis ball and told to see how quickly they can pass it around so that everyone touches the ball. At first they hand off the ball from one person to the next, or throw the ball at one another. Initially they may take as long as 20 to 30 seconds to get the ball to everyone. As the exercise progresses, the team learns to pass the ball around more and more efficiently—they drop it less, and the ball moves much faster. However, their time only improves 5 to 10 seconds. The group is then told to think about the purpose of the process—everyone touches the ball—and they are asked to be creative in the process they use. Most groups figure out a way to have everyone converge on the ball and finish the exercise in less than 5 seconds.

At first most groups are stuck in the mindset of passing the ball around. They worked to improve this process by passing the ball faster, in a continuous approach much like process management. When it is suggested that they think of new and innovative ways to accomplish the same end, they rapidly converge on a faster solution; they re-engineer the process.

This is not to say that process re-engineering is always the answer, and even when process re-engineering is selected, the end result is a process that you will need to place in a process management mode.

With process re-engineering it is important to consider the resources required to implement a re-engineering effort. Unlike process management, process re-engineering can consume a lot of money and resources in a very short time. The effort can be even more costly if outside consultants are engaged in the re-engineering work. We have seen process re-engineering efforts range from one person working full time for a few weeks at little expense, to a large team and a consultant with expenses in the millions of dollars.

Interestingly enough, many process re-engineering efforts can occur unknowingly. It may be happening in your organization right now, and you don't even know it. That's how it was with us. In 1985 we were involved in product development of a new telephone system for small businesses. Two engineers right out of school were given the responsibility for designing a new plastic housing for a next-generation telephone system. Plastics design had a reputation for being one of the gating factors for product development at that time. Intervals often exceeded six months just to get the first molded parts.

The traditional undocumented process went as follows:

> The engineer sketched the parts by hand or with a CAD tool, and took the drawings to drafting for entry into another CAD tool. The engineer

> would review the resulting prints (which were often quite different from the original sketches, since the drafting department members also acted as designers), and then made additional changes based on that review. The revised prints were then provided to a modeler who crafted handmade pieces (or sometimes molded them from a silicon-based rubber) for the engineer to review. These one-of-a-kind parts were sometimes used in the first working models to show the basic concept to others.
>
> After this handmade model was reviewed, changes were sent back to the drafting department to update the drawings. Once these final drawings were complete, they were then sent out to the manufacturing location, where two things would happen. First they were sent out to bid to potential toolmakers/molders. Next, they were re-entered into another CAD system, since the factory's CAD tool was different than R&D's.
>
> When the bidding process was complete, the selected vendor would begin the tooling process. Some six to twelve weeks later, the molder would send a sample part back to manufacturing, who would forward the part on to R&D for review. This process took almost six months to complete.

These two new engineers and their supervisor decided to change the process for their project. They skipped the in-house drafting, the in-house model making, and the bidding process for production molding. Instead, they designed a new process in real time. They began with some hand-drawn sketches. These sketches were shown to several molding vendors, along with a proposal to work directly with the vendor to both capture the design on the vendor's CAD tool, and make aluminum tooling.

The engineers picked a vendor, sat down with the vendor's CAD designers, and captured seven different plastic parts on the vendor's CAD tool in two weeks. This required that they work through every design detail with the vendor. The first molded samples were ready three weeks later. The process was high risk, and more expensive than the old process, but the savings in time were remarkable. In *five weeks* the first parts were available. Moreover, since the vendor was sharing responsibility for the design of the parts on their CAD system, they were more than anxious to make sure the parts fit together and that the designs were optimized for the tool-making and molding process. These aluminum tools were then transferred to manufacturing and were used for the first twelve to eighteen months of production.

This new process was not documented, and to our knowledge, not well understood outside a small circle of engineers. When the engineers left for other assignments, the process left with them. The old process was still being used some years later, and was still a major trouble area. The re-engineering had happened, but no mechanism existed for capturing, managing, and improving the new process for others to use.

This example probably repeats itself many times in many companies. Too much knowledge is lost when no process management structure is in place and employees move on to other assignments. The competency and past learnings within the organization rest wholly in the memory of the people. In large companies, this can result in one group performing at levels considered best-in-class for their industry, and a similar group doing the exact same work on the next floor suffering from quality and productivity problems.

In 1992 another team of engineers embarked on a re-engineering project for our high-level product realization process. They hired a consultant, and spent eight focused weeks designing the new process. With the consultant's help, they followed a very structured approach and produced a thick document describing the new process. Following this design phase, they initiated trials with several projects to test the new process and gather performance data.

During the trial period a process management team was established to implement and manage the new process on an on-going basis, with a budget exceeding one million dollars. This budget did not even include the consultant's fee nor the cost of the large design team for the eight-week period. Within eighteen months the implementation grew from five realization teams to twenty-one teams, with a target of forty teams within twenty-four months. The process management team re-engineered the documentation, developed customer-focused metrics, and designed training programs on the new process for the organization. The result of this effort was a true knowledge transfer from the original design team to the people who did new product development.

These two examples probably mark the extreme ends of process re-engineering, and most projects will fall somewhere in between. We have had the opportunity to support and be involved with many re-engineering efforts, and had the opportunity to observe others. We have concluded that successful re-engineering projects rely on some basic principles, and follow a process much like the process used for defining and developing a new product. This chapter attempts to capture these key principles, as well the steps for effective re-engineering.

PRINCIPLES AND PROCESS OF EFFECTIVE RE-ENGINEERING

A simple way to think about the re-engineering process is to:

- Start with a clean slate
- Begin with the customer in mind
- Employ "best practices"
- Deploy, deploy, deploy

To "start with a clean slate" means that you set the stage for open, creative thinking and bold new solutions. To "begin with the customer in mind" means that you first consider your customers' needs, wants, and requirements. To employ "best practices" means that you consider the best knowledge of your best people, and the best practices of competitors and noncompetitors. To "deploy, deploy, deploy" means that you don't stop until the new process is firmly entrenched and common practice within your community.

Successful re-engineering efforts are rooted in these principles. They allow out-of-the-box thinking, focus on customers, rely heavily on the best people in the organization, search out best practices, and don't quit until the new process is fully deployed.

You should place the best people you have available on a re-engineering task. These people should have a profound knowledge of the existing process and tools, understand the environment of process users, and be creative and resourceful. A few risk takers on the re-engineering team doesn't hurt.

The re-engineering team should agree upon a sound process for executing the re-engineering task, and this process should embody the principles discussed previously. The process model that most closely resembles the re-engineering process is the product realization process. The key steps of a product realization process can be modified slightly and applied for effective process re-engineering. These steps are:

1. Concept development and feasibility analysis
2. Planning and definition
3. Development
4. Verification
5. Deployment

STEP 1. CONCEPT DEVELOPMENT AND FEASIBILITY ANALYSIS

In most cases, a champion or project leader should be assigned to scope out the re-engineering effort and prepare a prospectus for the task. Because most re-engineering efforts are motivated by poor performance in development intervals, product quality levels, customer satisfaction, or other critical measures that demand a change, the project leader or champion has an initial responsibility to frame the problem.

This champion should begin by preparing a prospectus that includes:

- The scope of process to be worked
- A problem statement—a clear statement of the current state of affairs that captures the specific problems with the process today in terms of performance measurements, i.e., time, cost, and quality
- The impact the problem has on customer satisfaction and competitiveness
- A high-level description of the future state—a vision for how things would be after the re-engineering is complete
- The potential benefits (dollar, quality, or cost assessments) associated with a re-engineered process
- The estimated resources (number and type) required to complete the task (including deployment)
- A rough schedule for completing the major phases of the work
- A statement of potential risks
- A list of metrics to be used to assess effectiveness

This prospectus should be presented to management for review, and the project champion should not proceed until a commitment from management at the right level has been obtained. The "right level" means that sufficient management support is available to secure the necessary resources and budget to define and develop a solution successfully, and that the necessary management support will be available when the deployment phase begins.

Upon approval of the project prospectus, the project champion should assemble the re-engineering team. Any involvement with a consultant should begin at this early phase.

Team members should be dedicated to the re-engineering project full time. Re-engineering efforts are most effective when the team members can concentrate fully on the task at hand. The team may even want to go off-site to begin their work, thereby disassociating themselves from their previous assignments. It is very important that the team be allowed to come together and focus as a team. In some cases, specific team-building exercises may be employed to bring the team together before actually beginning the re-engineering effort.

The tasks of the re-engineering team during the concept phase include:

- Understanding the old process in detail, and establishing the root causes of the problems with this process
- Understanding who the customers of the process are, and documenting the needs of those customers
- Understanding what the needs of the business are for a new process
- Developing an external focus through benchmarking or with the help of consultants
- Establishing the fundamental principles for the new process
- Creating a team vision for how the new process environment will look when the re-engineering effort is complete

The team should fully understand the performance problems with the current process. This will usually require the team to work through the old process in detail to understand, as a group, how the process worked and where the problems were. "As a group" is an important requirement here, as most re-engineering teams are composed of experts in different areas, and they need to spend time together understanding the whole picture, not just their area of expertise.

The team should spend time flushing out the root causes of the symptoms evident in the process, and apply tools such as root cause analysis. When finished with this exercise, the team should be able to answer questions like What caused this process to take so long? or Why was rework or multiple iterations of the process necessary? or At what phase in the process were customer-found faults introduced? or Why is the process inefficient? or What elements of this process were dissatisfiers to our engineers or to our customers?

The team should identify the customers of the process, and understand (and document) the needs of those customers (recall the discussion in Chapter 4 regarding voiced vs. unvoiced customer needs). This effort will often require the team to go to the users and customers of the process and conduct interviews to get at the real needs of these customers. We have found that internal customers will initially state their needs in terms of what they currently receive (what they are used to getting), instead of what they need. Only through in-depth probing can you ascertain what they really need and want—what will truly satisfy them. We have found that people are very pleased to find out that you want to know what they *really* need to do their job.

A good example is a group of factory workers who were asked what information they needed for the assembly of an electronic housing. They initially replied that they needed an *A* drawing. An *A* drawing was a document that combined the entire assembly instructions into a single document. When asked to be more specific, it turned out that all they really needed was the parts list for each housing. They then observed that the current *A* drawing format took a lot of time to extract the data they needed, and on top of that, they were not always sure it was up-to-date. They often had to look at design change orders to see if any changes had been made. It was amazing to see the shock on their faces when they heard that we were trying to give them just the information they wanted, up-to-date, and in whatever format best suited their needs.

At the same time that customer needs are being identified, the team should be establishing a clear statement of the business needs the process will meet. These needs may be expressed in terms of reduced development intervals, higher productivity, or improved quality. Again, interviews may be necessary with the funders and key managers involved with the process to collect a full breadth of opinions of how the new process should perform to meet the needs of the business for lower costs, faster development times, and improved quality.

The team should also develop an external focus. They should learn about, or benchmark, other companies—possibly even competitors, to understand how others approach similar tasks. Consultants can play a unique role here because they bring the knowledge they have gained in working with other companies, including competitors. Their breadth of experience allows them to share the state-of-the-art practices of other companies without the need for extensive and time-consuming benchmarking by the team. If budget for consultants is not available, the team should plan to do benchmarking themselves.

The term *benchmarking* as used here refers to the process of learning about the practices of other companies, both competitors and noncompetitors, in your area of interest. In some cases you may be able to identify a best current practice based on some objective measures of how well different companies perform certain tasks or processes. We have found it difficult to ascribe this label to a particular process given the lack of comparable data and complete knowledge of all other processes. When we do use this term it is with the understanding that it is the best current practice that *we* know about. In other cases, you may only find elements of others' practices that are of special interest. We sometimes term these special findings *gold nuggets* to represent an idea or concept that, when combined with other ideas, would result in an improvement to our work.

Preparing for benchmarking, developing your benchmarking strategy, selecting your benchmarking sources and partners, and conducting effective benchmarking trips is an involved process not covered in detail here. Chapter 9 briefly summarizes some of our experiences with benchmarking, and we suggest reading works of several authors who have covered this topic in detail, including Robert C. Camp (Camp 1989, 75–118), Michael J. Spendolini (Spendolini 1992, 105–183), and Gregory H. Watson (Watson 1992, 51–76).

Another important step in concept development is the statement of the fundamental principles for the new process. These principles are identified throughout the research and benchmarking the team has done into best current practices, as well as from ideas within the team.

For a printed circuit board process, a set of concepts to be applied were:

- Concurrent engineering
- Cross-functional design teams
- Shared bill of materials
- Incremental design transfer to manufacturing

Another example is taken from a recent re-engineering team that focused on the high-level product realization process. The new process design was based on several principles that guided the re-engineering team through the definition process. These items were termed the *essence* of the new process, and included the following principles:

- Small, empowered, cross-functional teams own and manage the "offer" through the product realization process, and are accountable for the success of the product
- Customers are involved throughout the definition, development, and verification phases
- Fundamental invention during development is minimized through the use of technology platforms

- Development of the supporting elements of the product (servicing, documentation, training, and marketing) is done concurrently with the product development
- Quality gates are established in the process where requirements (criteria) are known and met

Principles like these guide decision making throughout the definition and development phases, and are useful in teaching the new process to others during the deployment phase.

Once the team has worked through these concept development steps, they should be naturally developing a vision for how the new process and tool environment should look. They should be able to describe this future state, and be able to set high-level performance targets. These objectives may address interval (e.g., 40 percent reduction in time-to-market), or quality (factor of ten reduction in defects in five years), or any of the other performance areas targeted by the re-engineering effort.

The output of the feasibility and concept phase should be:

- A project prospectus
- A fully engaged re-engineering team
- A needs statement and concept description

The needs statement and concept description should include:

- Root causes of current process problems
- Customer needs
- Business needs for new process
- Fundamental principles (best practices) that the process is to be built upon
- A vision for the new environment, including high-level targets for performance

The needs statement and concept description should contain all the elements discussed here and be sufficiently detailed to allow the team to begin the process of writing requirements and creating an architecture for the new process.

STEP 2. PLANNING AND DEFINITION

The definition phase for re-engineering a process will be the most challenging for the team. The purpose of this phase is to architect a new process, beginning at a high level and ending with each detail, and to write specific requirements for the process and related tools.

One re-engineering team we worked with elected to travel off-site for several focused days to work through a new process and tool architecture for the process. The message here is that process re-design for major processes requires prolonged concentration on the part of the whole team. The team should wrangle with many of the potential solutions and work toward that optimum architecture.

Critical to the team's success is their ability to work effectively as a team. We have found it useful to engage team facilitators to help with team building, and professionals in process re-engineering to help with the re-engineering process. One thing you can ill afford is to spend time working on team issues and personality conflicts later in the re-engineering process. These things are best handled up front when the team is just getting together.

The new process should be flowcharted from beginning to end as the team works through every process step, interface, and tool interaction. Each step should be evaluated and the question "Is this a value-adding step?" should be asked. All non-value-adding elements should be evaluated for other alternatives.

It is useful to begin this process by identifying the core elements of the process. Almost all processes have high-level steps that are central to the process. As an example, take circuit board development. At a high level, you can identify the following steps as core elements:

- Hardware specification (requirements)
- Schematic capture (electrical design)
- Printed wiring board layout (physical design)
- Manufacturing process and test development (verification)
- First build
- Unit test and integration testing
- Final documentation approval

Surrounding these core elements are the multitude of activities (process steps) and tool interactions that are necessary to develop the product. The re-engineering team should define these process steps to align with the principles developed in the concept phase. The result should be a detailed process definition that includes a diagram of the information flow accounting for every document and piece of data that will go through the system.

The process description should also describe the roles of the process users—the *who*—associated with each step. Recall the discussion of "Roles and Responsibilities of Process Users" in Chapter 4. When re-engineering a high-level product development process, the definition of these roles and responsibilities of people and teams may require even more of the team's effort than defining the process steps.

Finally, the tool architecture (i.e., systems, software applications, etc.) should be described, with all key interfaces identified and understood. In each of these areas the re-engineering team should strive for simplicity. The most elegant solution will most likely be the most simple as well.

Following this effort, the team should be able to document the new process architecture, team structure, and tool environment, and then write requirements for the process and tools that will require development. The tool aspects of a new process will typically entail more requirement writing than the process itself.

These requirements should also include the specific needs for documentation and training. Along with management support, good training and documentation will be the cornerstone for a successful deployment of the new process.

A common pitfall for re-engineering efforts at this definition phase is to base the process redesign around a tool. Although tools in many ways impact the process, this phase should focus on customer needs and required steps of the process, and specify the requirements for the tools (not vice versa). If the tool becomes the driver for the process, the team may fall into the trap of automating a poor process. This just results in producing more garbage faster. Computers are great tools, but should be viewed as enablers for making a good process better and faster.

The team should also develop a detailed plan for the development, verification, and deployment of the process and tools. This plan should include:

- Updated resource requirements
- Expense or capital needs
- Detailed work plans and schedules for each subsequent phase, including development, verification, and deployment

This new plan should be reviewed with the management sponsors to re-affirm their commitment and to obtain the necessary resources and budget for the rest of the project.

The output of this phase will be:

- Detailed requirements for tools, including a description of the tool architecture and interfaces with other tools
- High-level architecture for the new process
- Detailed process designs that show the process flow, information flow, and roles of process users
- Requirements for documentation and training on the new process and tools
- Detailed plan for completing the development and verification work

STEP 3. DEVELOPMENT

Development for re-engineering processes will focus on one or more of these areas:

- Process documentation
- Training on process and tools
- Tools and tool interfaces

When new tools are part of the re-engineering effort, tool development is usually the most well understood area for a product development organization, and requires little discussion here. Most engineers in R&D are accustomed to the development of tools involving hardware and software, and do this quite naturally.

The development of documentation and training, on the other hand, is quite often ignored by engineers, almost as if it is a nuisance. For process re-engineering, training and documentation are crucial to the success of the effort. Documentation should be short, simple to understand, and should highlight the essential elements of the process. Engineers do not want, nor will they read, inches and inches of process documentation. Most of the successful process managers that we have worked with have had short, easy-to-understand documentation. Those that have been less successful often point to a thick ring binder that is now a nice, permanent addition to everyone's book shelf.

Documentation should rely heavily on the elements for good process descriptions discussed in Chapter 4, including a purpose statement, clear inputs and outputs with supplier and customers identified, key process steps described, templates for key documents, checklists that summarize the important steps that should have been completed, and clear criteria for moving from one phase of the process to the next (quality gates).

Documentation for tools should also be short, relying on tip sheets, sample screens and input, and an easy-to-use reference guide for the features of the tool.

Training should be based on the documentation, and should avoid creating yet another set of documents. Training should include a historical account of the re-engineering effort, the motivators for change, the benefits associated with the new process, and an engaging presentation of the desired end-state—the team's vision for how the new processes and tools will work. Training should also be based on enabling objectives. An enabling objective can be phrased "At the end of this course, you will be able to . . . "

One re-engineering team we studied developed an overhead presentation based on their design work. Their first attempt at training was a two-day session where different members of the re-engineering team presenting their overheads. Four teams were subjected to this training, which quickly got the reputation for being a disorganized series of overhead presentations—a flop. Later the same material was condensed into a one-day session, still in an overhead presentation format. This time a professional trainer helped coach the presenters and improved the organization of the training. Even after this, the feedback from teams included statements like "boring," "too much information," "just a core dump," and "I left without knowing where to start."

A full twelve months passed before the team realized that the quality of the training had a direct impact on the quality of the deployment effort, and they had to start over and develop a meaningful and effective training program. Working with skilled training designers, they developed a new course that included many simulations and scenarios, actively involved the students in the learning experience, and utilized testimonials from real-life projects. This new training focused on enabling objectives, and led to an effective training experience.

The output of the development phase should include:

- Completed tool development (including interfaces with other existing tools)

- Process and tool documentation
- Process and tool training courseware

STEP 4. VERIFICATION

Every new process, tool, and training course should be verified with trials. This will give the re-engineering team an opportunity to work out any bugs or kinks in the system, and understand the impact the new process will have on the user community. This does not mean that all testing activities should wait until development is complete. Many subelements of the process, systems, and tools should be tested concurrently with the overall development. What is required is that the larger, integrated elements of tools, processes, team structures, etc., be verified. The most important aspect of trials, second only to surfacing bugs in tools or identifying weaknesses in the process, is assessing the impact of a re-engineered process and tool on the users.

We were involved in a re-engineering effort with a factory that was served by two development locations. This effort included many new processes for design information transfer into manufacturing, material management, and change management. The re-engineering team was also introducing a new tool to help manage the design and manufacturing information for the new processes.

The impact on the user community for this project, especially product engineering, was significant. Many engineers felt overwhelmed by the learning curve associated with the new tool, and felt burdened by the need to learn something new while still trying to get their current work load processed. The results of the early trials caused the re-engineering team to craft a deployment program that would be sensitive to the high impact the new process had on the community.

Unlike most manufacturing processes, the intervals for product realization processes are too long for the team to implement a trial program and then wait for the project to complete. For example, a new plastics development process may take six months to execute the process from end to end, and a new product realization process for an electronic product may be six to eighteen months for a full trial. Remember that re-engineering was selected in the first place to get real gains in performance quickly, and verification needs to account for this need to move ahead with reasonable speed.

As a result, the new processes and tools will need to be verified by simulations and by piloting elements of the new process with different projects. The team should clearly plan what elements of the new process or tools require testing, and develop a verification plan for each.

These verification plans should include a feedback mechanism for process users. The team should react quickly to those items that are having a negative impact for the process users. The reputation of the effort can be quickly scuttled if the community itself talks the project down. This can be avoided by collecting feedback from the trial projects and demonstrating pro-active efforts to correct these outages quickly.

STEP 5. DEPLOYMENT

Once the key elements of the process and tools are verified, a deployment plan should be developed and initiated. This deployment plan should include the on-going process management of the new process. If the process has an impact on many users and many projects, then a dedicated implementation team may be needed.

Deployment of most new processes is underestimated by many re-engineering teams. They fall short on forecasting the time and expense needed to implement a new process or tool. The factor with deployment most underestimated by re-engineers is the impact of changing from the current process to the new process. The second most underestimated factor is the resistance from the user community to change. The re-engineering team can easily develop an attitude regarding their new process that can be perceived as pushy and arrogant by the users, who may not have a vested interest in changing anyway.

Deployment above all requires persistence and patience. Planning should be conservative, and should account for transition issues up front when possible.

As discussed in Chapter 4, successful process deployments rely on several key elements:

- Clear communication of benefits the process brings, both to the organization and to the users
- Effective training and easy-to-use process documentation
- Frequent use of exit/entry criteria and quality gates in the process
- Management support for the project clearly communicated to the user community

As we mentioned previously, the project leader or champion should not have formed the re-engineering team at all if the prospectus was not approved and supported by top management. The reason for this becomes evident during deployment, when the support of management is essential for the successful deployment of a new process. This support cannot be underestimated. We found that even a letter from a top manager expressing their expectations of the community to use the new process can make a large difference.

Usually, however, a letter is not enough. For example, we worked with a process management team (PMT) that was chartered to deploy a process to six different organizations. Each organization performed a similar function for different market segments. After twelve months of deployment effort by the PMT, the level of implementation ranged from none to role model. When we looked further for the cause of this difference, we found that in the organization that had little or no implementation, there was almost no management support. Moreover, some of the managers even told their people not to let the process get in the way of their "real" work. In the organization that was viewed as the role model, strong management support was evident with middle managers. One manager even told their folks, "this is it [the process], let's work together to make it work."

Easy-to-follow documentation and training is important for obvious reasons. The user community must develop an understanding of the process and be able to apply the process without too much pain. If the pain of learning the process is too high, many will simply ignore it. That means that the process documentation must be short, accurate, and of high quality. It must be easy to access and use. Training must be engaging and focus on enabling objectives such as "When you finish this course you will be able to . . . " as discussed previously.

Clearly expressed benefits help the users understand why the change is being made. Again, the re-engineering prospectus will be valuable to recall the key problem areas and performance outages. The re-engineering team should have re-evaluated the benefits associated with the new process and be able to relate them to the original problems and performance measures. This should be integral to the training program. Helping people understand why the change is being made goes a long way to lowering their resistance factor.

Most of all the deployment phase should establish a mechanism for managing the new process on an on-going basis. In some cases a process management team may be needed to manage implementation and on-going improvements as described in the process management process in Chapter 4. In other cases it may suffice to have one engineer play the role of process owner and manage the process on an on-going basis.

The re-engineering team should approach deployment with as much energy and planning as they did the design of the process. They should document the deployment steps and manage the deployment as they would manage any change to the organization. Deployment activities of senior and functional managers, the deployment team, and the users should be identified, and the team should carefully manage the execution of these activities. The re-engineering team should view the change to the new process as a transition for the organization that will take time and energy, and not an event that happens overnight.

Senior and functional managers will need to serve as sponsors or champions of the change both while it is being introduced and during the transition. In order for them to carry out that role, the re-engineering team will need to help them understand the change and gain their support. The management in turn will need to sponsor the change to the organization. With this management support and pro-active sponsorship, you will be well on your way to a successful change to your new re-engineered process.

6 Principle-Based Problem Solving

Strategic Quality Ideals

✓ Focus on Process: the How of Doing Work

In Chapters 4 and 5, we have described two primary methods for process improvement. In Chapter 4, we described an approach for continuous, incremental improvement to establish working processes. In Chapter 5, we talked about the methods we have found useful when we had to make radical changes or when we needed breakthrough improvements. In both of these tasks, we have found problem solving to be an important cornerstone for process improvement. In this chapter, we talk about how we've gone about working with our teams to give them the skills—the tools—they need for that task.

Rather than just describe a number of problem-solving tools, or techniques, that we've used for tackling process improvement, we take a slightly different approach. First we develop a rationale for our approach, then we develop our principles for effective problem solving—an approach that relies on understanding and common sense, rather than applying a complicated series of steps performed by rote.

A RATIONALE FOR EFFECTIVE PROBLEM SOLVING

Have you ever gone to a meeting where ideas were criticized and batted around? It usually takes just once for an idea to be publicly criticized for some people to learn to just sit there and be quiet. Often one or two people are able to control an entire meeting simply because they have the most forceful personalities. In these cases, we lose the contributions and experiences of the others on the team—something we can ill afford. We need the group to work together to achieve their maximum potential in solving these often difficult improvement problems.

Problem-solving tools can provide structure and openness to what otherwise can be a very uncomfortable environment. These tools can help us to look past the symptoms, find the root cause of the problem, and work these largest drivers first. Most of all, they can help us work together in a productive team environment.

In many cases engineers learn problem-solving tools one at a time, in isolation from an entire problem-solving process. The result is an inability to apply more than one or two tools in any given situation. Groups we have observed working on process improvement use brainstorming more than any other tool. Running a close second is root cause analysis. Beyond these two commonly applied methods, we seldom see other problem-solving tools used. Moreover, groups usually begin their problem-solving effort using one or two tools at their first meeting, and then "play it by ear" (i.e., anything goes).

In other cases, engineers learn an entire problem-solving process with all of the associated tools, and have the opportunity to apply the process only once in a while. As time passes, the skill of applying the process and tools effectively is lost.

We have had the opportunity to teach individual problem-solving tools as well as to teach an entire process for applying these tools. The best- known example of a full problem-solving process is the Quality Improvement Story, QIS.[1] This process, which we've included for reference in the Appendix, guides you through a prescribed set of steps that clearly define the problem, establish the root causes, evaluate the largest contributors using a Pareto method, and develop counter measures.

The difficulty with methods like QIS is that people have to remember the steps, and, for infrequent users of this tool, that's hard—especially if you want to sustain an effective problem-solving environment over a long period of time. We have even observed quality experts faced with problems in their own field who disregard problem-solving processes in their rush to solve the problem. We have facilitated many groups through the standard problem-solving tools, and, even with facilitation, people struggle.

So what can we do? One alternative is to capitalize on the fact that people can remember and apply the principles behind good problem solving much more easily than they can remember and apply a long series of tools.

1. Developed and copyrighted by Florida Power and Light, Inc., to help them in their quest for the Deming Prize. FP&L won the award in 1989.

For example, we have learned the importance of not working the symptom of a problem, recognizing instead the need to understand the root cause. However, when a group uses a tool like root cause analysis, they often forget a follow-up step called symptoms analysis. We have observed many root cause diagrams where many of the items are symptoms. That's because it is easy to forget the intent of the tool, and to just go through the motions mechanically. This is not unlike asking someone to do something without telling them the reason behind your request. In many cases they may do exactly what you say, but not at all what you meant.

We believe the challenge is to teach the principles, the purpose behind the tools. Then a team can use the tools to apply these principles effectively. When the group forgets the tool, then they can often improvise. This position may alarm some people, but we would much rather see a group work real root causes, use data to assess the largest drivers, and openly involve everyone in the solution design with no pictures or tools whatsoever, rather than see a technique rigorously applied without adequate understanding—which can lead to incorrect conclusions. In most cases it will not matter if the group designs a new tool in real time, as long as it is based on the right principle. For example, take a Pareto diagram. This is nothing more than a tool designed to identify and illustrate the biggest drivers. Does it really matter if you draw a bar chart or not?

THE FIVE PRINCIPLES BEHIND EFFECTIVE PROBLEM SOLVING

With this said, does that mean problem-solving tools are unimportant? Not at all. Tools like the Pareto charts and fishbone diagrams show data in a common format that makes communicating your work to others easy. Other tools do the same thing, and, when applied correctly, help groups work through problems quickly and productively.

We believe that tools are easily remembered when they are easy to use and based on commonsense principles. The principles behind effective problem solving are:

- Clearly defined problem
- Participation without criticism
- Addressing root causes, not symptoms
- Prioritization of causes and solutions
- Data-driven decisions

CLEARLY DEFINED PROBLEMS

Consider the first principle, "clearly defined problems." In most cases, no tool is even required for this step. The group just needs to agree on the problem statement. We are surprised the number of times a group will begin working a problem without ensuring that each person has the same idea of what the problem is. In addition to a problem statement, it is helpful to note the indicators or measures that made this problem visible in the first place.

A recent problem-solving exercise we supported focused on ineffective product realization processes for international developments. What was helpful to know up front, before tackling this problem, was that the indicators were slow time-to-market, churn in requirements, and poor quality of product delivered to customers. The single indicator, however, that brought this problem to the forefront was the recognition that expenses for supporting these development efforts were far above the benchmarks. Given the problem statement and understanding of these indicators, the group was able to begin addressing this problem effectively.

PARTICIPATION WITHOUT CRITICISM

Now consider the concept of "participation without criticism." The following tools support this principle:

- Brainstorming
- Affinity Diagrams
- Multi-voting

Brainstorming

The purpose of brainstorming is to generate ideas. When a group begins brainstorming, they are told that all ideas will be written on an easel or a blackboard without evaluation or criticism during the first round. It usually begins with someone saying:

> *Let's **brainstorm** some ideas. That means that no matter what anyone says, we just write it down. We won't discuss it, criticize it, or even ask what they meant. We just write it down. After the group runs out of thoughts, we will go back and discuss each idea in more detail.*

After correcting one or two people on this rule, the process usually works well and gets everyone involved. This first round focuses on quantity, not quality, of the ideas, and works to produce a "free wheeling" environment (Osborn 1963, 151). The second round through the ideas allows the group to clarify the ideas, eliminate redundant items, and combine common items. The output of brainstorming is a list of ideas. For most groups, brainstorming need be no more complicated that what is described in this short section, and fully supports the principle of participation without criticism.

Affinity Diagrams

The purpose of affinity diagrams is to categorize or group items from a random list, i.e., to group items together that have some *affinity*. The process is also simple. Write every item down on a Post-it note. If you have twenty ideas, then you will need twenty Post-it notes. Then put all of the Post-its on a wall surface or white board. The group gathers around the Post-its and someone says,

> *The purpose of this exercise is to identify natural groupings of items by silently and simultaneously, everyone working at the same time, placing the Post-it notes with other Post-its that belong together. No discussion is allowed. Anyone can continue to move Post-its around until everyone is content with the groupings (or tired, whichever comes first).*

This is a silent exercise. Once the group has finished this silent grouping, then everyone discusses a name for each category. The output of an Affinity Diagram is a chart such as shown in Figure 6.1. The titles for each box are the category names selected after the Affinity silent exercise. This tool usually requires 15–30 seconds to explain, and again fully supports the concept of participation without criticism.

This simple example is not untypical of the output of a session to generate an Affinity Diagram, and it demonstrates one of the problems we often encounter when we use this technique (see also Brassard 1989, 23). Frequently we opt for

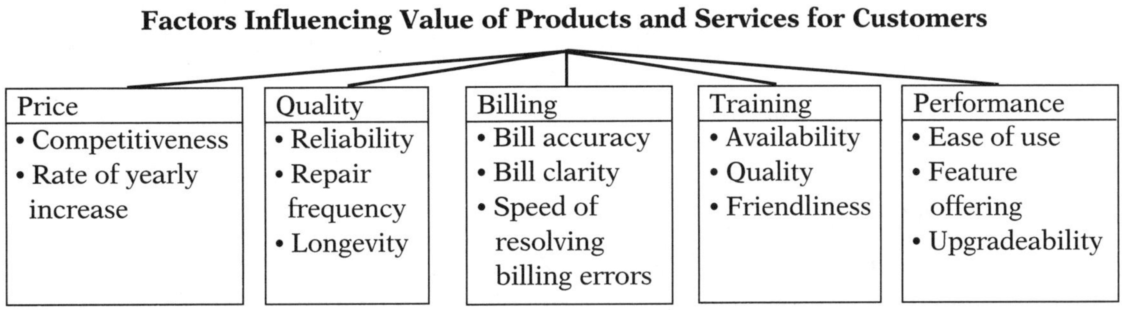

FIGURE 6.1 *Affinity Diagram*

brevity, using simple one- or two-word descriptions, when, in reality, we've obscured our meaning with our shorthand. For example, we've used "Competitiveness" under the "Price" heading, when what we mean is that "our price is higher than our competitors'." We strongly recommend avoiding brevity in favor of being more precise and complete to ensure that your meaning is not lost, in the following discussion and later, when you are trying to document your work clearly. One technique that can help achieve this completeness is to ensure that the description includes a verb as well as a noun.

Multi-voting

Consider another example—multi-voting. The purpose of multi-voting is to prioritize a set of items, or select a single item from a list. The process begins with a random list of items (say 20 for this example), typically written on a flip chart or white board. The process starts with someone in the group saying,

> *Why don't we multi-vote this list to reduce the number of items we are working with? We will do three rounds. Each person in the group will get five votes the first round, three votes the second round, and one vote the third round. In the first round, each person can vote for five items (one vote per item).*

After giving the group a few minutes to select their items, one at a time each person places a mark by the five items they selected. After everyone has marked their five choices, then the number of votes per item is tallied. The top half of the items are the new list for the second round. So if you started with 20 items, then the top 10 are selected for round two. For round two, each person gets three votes. After the scores are tallied from round two, then select the top five, and do round three. For round three, each person gets one vote. The result is a prioritized list of items based on equal input from the group.

Note that all three of these tools allowed open and equal participation of all group members. Ideas were surfaced without criticism in brainstorming, ideas were categorized silently with the Affinity exercise, and voting input was equal in the multi-voting process. Each of these tools is designed to bring equity and fairness to group interactions.

ADDRESSING ROOT CAUSES, NOT SYMPTOMS

Consider the principle of "addressing root causes, not symptoms." The following tools support this:

- Brainstorming
- Affinity Diagrams
- Symptom analysis
- Root cause diagrams
- Interrelationship Digraphs
- Repeated questioning

Symptom Analysis

We've found brainstorming and Affinity Diagrams, discussed in the preceding section, useful in generating potential root causes and categorizing them. *Potential* is a key word here. Many items that surface in brainstorming are not root causes. They are merely symptoms of some other cause or causes. One way to determine whether an item is a cause or symptom is to ask the group to consider one item at a time and ask the question, What are the possible causes of this item (assuming first that it is a symptom)? Another method is go down the list and have the group assign an *S* or *R* next to each item to indicate whether it is a symptom or root cause. Then go back to the item marked with an *S* and evaluate potential causes for that item.

Root Cause Diagrams

The term *root cause diagram* is one of many used to describe an effective visual technique for displaying the factors leading to a particular result. These diagrams are also referred to as *fishbone, cause and effect,* and *Ishikawa diagrams*—*fishbone* because of the way they look, *cause and effect* because of the relationship they display, and *Ishikawa* after Kaoru Ishikawa, who first described the approach in the 1970s in Japan (Ishikawa 1982).

Construction of the diagram is most easily described by example based on the Affinity Diagram, Figure 6.1. The result, or effect, is the "customer's perceived value." The factors leading to that performance, the major categories, are the "*bones*." The subcategories, those items shown under each heading in the Affinity Diagram, make up the next layer of bones. The resulting root cause diagram is shown in Figure 6-2. While additional layers of bones can be added, we've seldom found it advantageous to add more than three layers of bones off the horizontal; the additional complexity more often obscures, rather than clarifies the problem at hand.

We've found the fishbone, or root cause diagram to be most useful for displaying, rather than discovering root causes. Our associates are more likely to

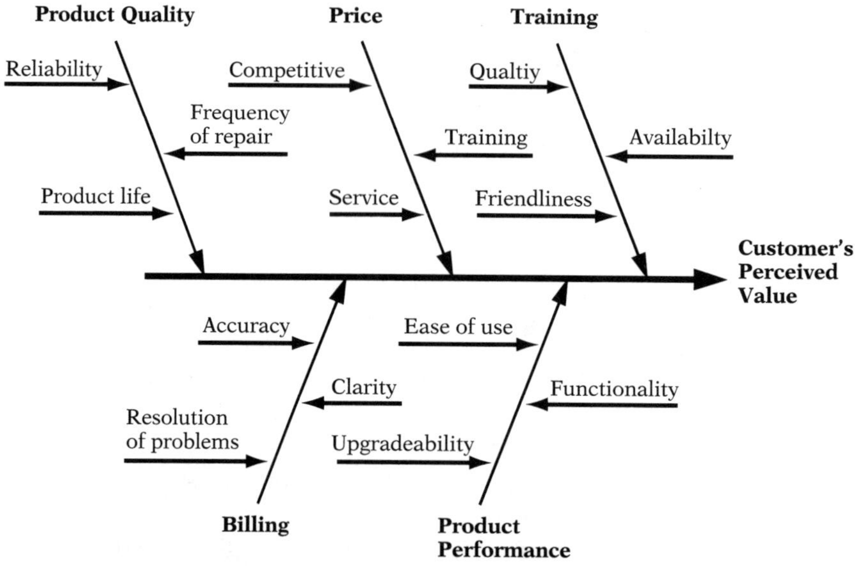

FIGURE 6.2 *Root Cause Diagram: Factors Influencing Value of Products and Services to Customers*

use another approach, depending on the type of information they have, to find root causes.

Interrelationship Digraphs

Interrelationship Digraphs provide a method for determining the relationship between a large number of items and can often be effectively used to determine the root cause of a problem, even when hard data are not available.

The essence of the technique is for a group to examine the relationship of the individual items through an orderly, pair-wise comparison. This is done graphically (hence, the name). The items can be arrived at through any of a number of approaches. The approach generally depends on the type of problem you're trying to solve and the type of data, knowledge, or experience on which the work is based. The results of a brainstorming session or an Affinity Diagram are often the starting points—although we've usually used a brainstorming session to generate an Affinity Diagram. We've then used the categories from the Affinity Diagram for the Digraph.

First, we have collected our information into a number of groups, such as we'd have from an Affinity Diagram exercise. Note: we've found that fewer than five groups tend to lead to trivial results; more than ten begins to become un-

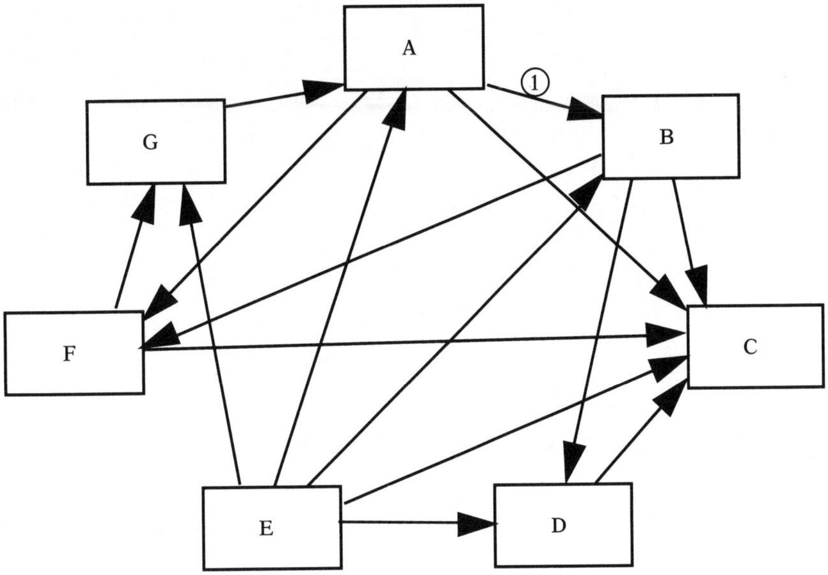

FIGURE 6.3 *Interrelationship Digraph Construction*

manageable.[2] We then arrange them on a board or easel, in a circle if we have enough space. We then begin to examine the relationship between the items by asking the question, "Does the first item cause the second to happen? Or does it influence the second to occur?"

If the answer is yes, we draw an arrow from the item that is the cause to the one that is influenced. If there is no relationship, we do not draw an arrow. For example, if we have seven groups labeled A through G, we can begin by comparing A and B. In this case A can cause B, so that A is a cause and B an effect, so we draw an arrow from A to B. We've labeled this arrow ① in the figure.

We continue the pair-wise comparisons, one at a time, comparing A with C, A with D, and so on around the circle in Figure 6.3 through G. Then we start the comparisons of B with the others, starting with C, then D and so on (note that we've already compared B with A).

When we have completed our examination of all the pairs, we can then calculate a "score" for each item. The score is equal to the number of arrows (i.e.,

2. For a larger number of groups, we found it beneficial to combine several items in an Affinity exercise to reduce the total to a more manageable number of groups. Using such an approach, we have seen the technique successfully used for a very large number of individual items (>400).

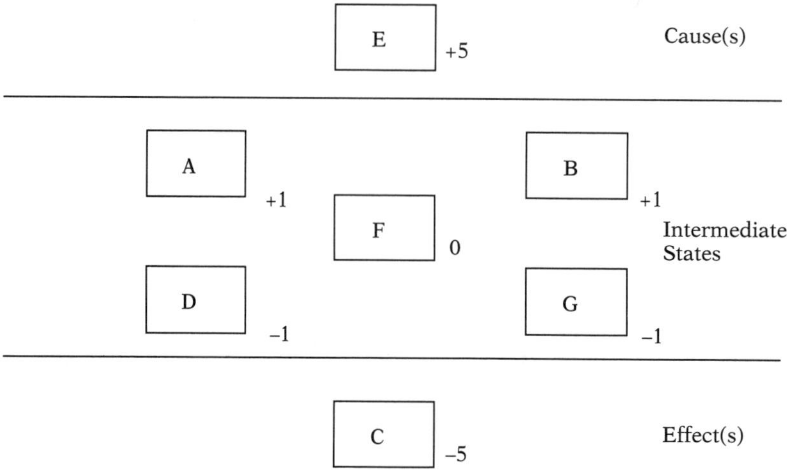

FIGURE 6.4 *Interrelationship Digraph, Redrawn to Emphasize Results*

the number of causes) leaving the group minus the number of arrows entering the group (the effects). For seven items, as in the example of Figure 6.3, the maximum positive number is six and the maximum negative is minus six. Further, the items with the positive scores are candidates for root causes; those that are most negative, the effects. Those with intermediates scores are transition states. For the example in Figure 6.3, the score for A is +1 (3 minus 2); for B, +1; C, –5; D, –1, and so forth.

We've redrawn the somewhat complicated Figure 6.3 without the arrows, but with the scores, for clarity in Figure 6.4. Further, we've gathered the groups into three categories, Causes, Intermediate States, and Effects. The Causes have the highest (positive) scores and the Effects, the lowest (negative) scores. In this simple example, we have only a single Cause and Effect, although we often have more than one of each.

The method for making the pair-wise comparisons and the determining root causes and effects is mechanical. We've often found it hard for groups to agree on "which way the arrow goes" when making the pair-wise comparisons. Remember, the arrow goes from the *cause* to the *effect*.

We've been generally satisfied with the results—provided we tested them to see if they, in fact, made sense. We've also found we need to parse the result into items within the span of the work at hand or, stated another way, within the scope of the proposed effort. Only those items within the scope of the effort need to be addressed in the subsequent work.

Repeated Questioning

Repeated questioning is our name for a technique for establishing root causes that relies on repeatedly asking a series of simple questions about a problem that has been uncovered. The questions are simple and direct:

- When was the problem found?
- Where could the problem have been found?
- Where should the problem have been found?

The first question is simply a statement of fact, where we discovered the problem. The second question marks the beginning of our investigation to determine the root cause of the problem. We ask the question at each prior stage in the process in an attempt to determine where we could have found the problem earlier in our development. When we reached the point where we can't push the problem back any earlier in the process, we conclude we've reached the root cause; we've found when we should have discovered the problem, using an appropriate technique or screen.

The answer to the first question is obvious and it sets the stage for the next question, Where could the problem have been found? At this step, we have to ask a series of questions to discover whether we have found a root cause. For example, let us consider the case of possible impact of a screen, or test, in surfacing the problem at this stage.[3] There are two cases, depending on whether or not a screen could be effective. If a screen would not have been effective, then clearly we haven't found the root cause and we continue our search.

If, however, a screen could have been effective, we have two cases, depending on whether or not a screen is currently in place. If, when we ask, Was there a screen?, the answer is no, the root cause *may* be that the lack of a screen is the root cause. If, however, the answer is yes, we have to ask, Was the screen adequate? Then the root cause *may* be that the screen was inadequate. Just because we have found a possible cause, however, doesn't mean we have found the root cause.[4] Further, we may want to consider the potential process improvement by improving the screen.

Our search does not end here. We need to continue to evaluate the upstream functions to see if the real root cause can be pushed to an earlier point in the development process. We do this by asking the third question, Where should

3. Adapted from an internal AT&T publication on root cause analysis.
4. See (Jones 1985) and (Grady 1989) for discussion and some detailed examples of how this technique has been applied in a software development environment.

we have found the problem? Essentially, we then repeat the earlier questions until we are satisfied that we have found the earliest point in our process that we can implement an effective screen for the problem.

A key point to note is that we've found that the techniques works well when the work is following a repeatable process (documented or not). The effectiveness of this technique degrades quickly as the variability of the process increases from one time to the next. We've personally seen dramatic process improvements (interval improvement of 30 to 50 percent—while also improving the quality of the output) when the technique is rigorously followed. We'll give an example later in this chapter.

PRIORITIZATION OF CAUSES AND SOLUTIONS

Consider the principle of "prioritization of causes and solutions." The need for this task is obvious to any group who ends up with 30 root causes or 20 potential solutions to a problem. The tools that support prioritization include:

- Interrelationship Digraphs
- Multi-voting
- Pareto charts
- Prioritization worksheets

Interrelationship digraphs and multi-voting are techniques reserved for situations that lack hard data to drive the prioritization. As previously discussed, they rely on the collective experience of the group.

Pareto Charts

Nearly forty years ago, Juran described the principle of separating the "vital few" and the "trivial many" to determine priorities in managing problems (see, for example, Juran 1964, 44). He called this the Pareto principle.[5] Sometimes this technique is also referred to as the *80/20 rule,* that is, the principle that 80 percent of the problems/issues/. . . are due to 20 percent of the causes. We're interested in the principle because of the insight it provides in analyzing problems where we need to determine priorities. Essentially the technique relies on gathering data on the frequency of occurrence of the various param-

5. See Pareto Principle in the glossary for an interesting account of the origin of this term.

TABLE 6.1 *Distribution Data for Pareto Example*

Parameter	Occurrence	Frequency (%)
A	180	45
B	112	28
C	48	12
D	36	9
E	24	6

eters of interest, and then graphically displaying them in a way that highlights the "vital few." The resulting graph is called a *Pareto chart.*

Stated another way, a Pareto chart is a simple graphical technique for displaying the relative importance of problems or issues. Figure 6.5 is the Pareto chart for five different parameters shown in Table 6.1. The frequency with which each parameter occurs is displayed as a vertical bar in percent against one axis, beginning with the problem having the highest likelihood. Often the cumulative fraction percent is plotted against the other axis, as in Figure 6.5.

Prioritization Worksheets

Prioritization worksheets, or prioritization matrices, provide an orderly way of categorizing issues, problems, and work based on weighted criteria, where the criteria are known or developed by the group. While relatively straightforward to use, we've seen them used only occasionally by our teams. They tend to find the process for generating the worksheets too mechanical and time consuming. (See Brassard 1989, for several examples of how these can be

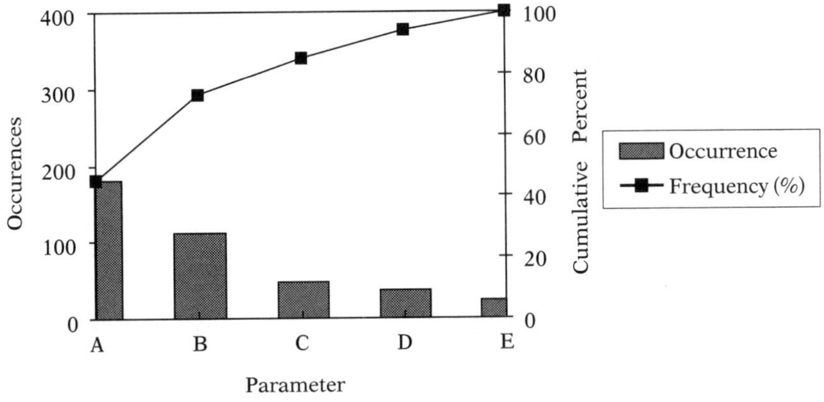

FIGURE 6.5 *Pareto Chart: An Example*

used to help prioritize tasks, issues, or possible options based on known, weighted criteria.)

DATA-DRIVEN DECISIONS

The last principle, "data-driven decisions," is a good rule of thumb for any problem-solving team. Decisions based on data nearly always drive better and more effective changes. Our associates seem to be much more willing to work with data; these facts seem to reduce the "politics" and the emotion associated with changes. Many tools support the presentation of data that we will not cover in detail here; they can be found in several of the references. Those that we've found to be among the more useful include:

- Pareto charts
- Run charts
- Control charts
- Check sheets

Run Charts

One of the simplest techniques we've used is a run chart. A run chart allows us to see how a parameter changes as time progresses. The run chart simply plots the data against time. The value of the data is shown on the y, or vertical, axis and time is shown on the x, or horizontal, axis. An example is shown in Figure 6.6. We'll show how we've used it as part of an example later on in the chapter.

Control Charts

A control chart is used to display how a (critical) parameter, measurement, or event varies with time. Hence, a control chart is a specific type of run chart. Shewhart is usually credited with developing this approach for analyzing the statistical variation of parameters observed during manufacturing. This technique has been developed to a high degree of sophistication, with particular techniques for handling various types of statistical variations, for example (see Shewhart 1986). We can then use the control chart, with precisely defined upper and lower control limits to help us analyze performance. Several examples of control chart look like the run chart shown in Figure 6.6. Kume 1985, Chapter 7 provides details for constructing such diagrams. AT&T is devoted to applying the techniques to manufacturing (AT&T 1956).

We've seldom used control charts in R&D. We're more likely to use these techniques to study problems such as we might find in production, where statistical variations in our data become important.

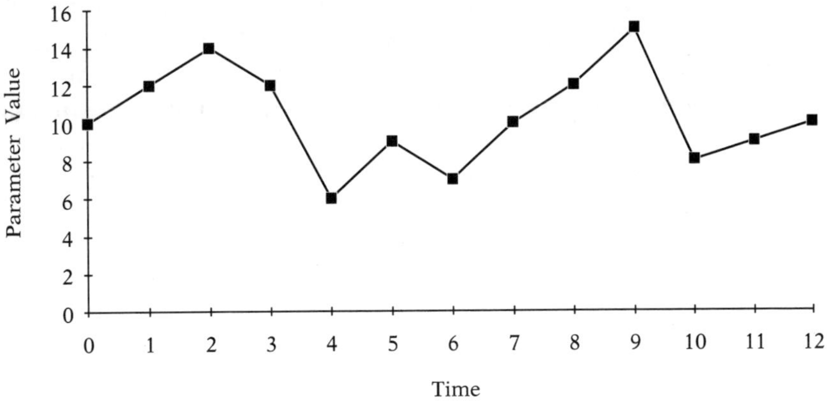

FIGURE 6.6 *Run Chart: An Example*

Check Sheets

A check sheet is nothing more than a table used for gathering data in an organized way for use in analysis and in looking for trends. We mention it here because it is simple to use, and you may read about it or see it being used. In practice, we would probably gather the data used in the run chart example presented here by using a check sheet. We'd then summarize those data in a table like Table 6.1, which we'd then use in our problem analysis. Kume (1985, pf, 10) provides several examples.

SOME COMMENTS ON TOOLS

We've described a number of tools for tackling process problems. The tools are basically of two types—those that depend on data and those that are based on knowledge or experience. Those that are based on data are often referred to in the literature as the *Seven QC Tools* or *Seven Basic Tools* (see, for example, Bakerjian 1993, 9–10, and Brassard 1989, 7 and 261). They are flow charts, Pareto charts, root cause diagrams, run charts, histograms, scatter diagrams, and control charts.[6] We've found the first four of these to be useful in R&D. For a more detailed discussion of these and the remaining Seven QC Tools, see Brassard 1989, Appendix B.

Those tools that rely on knowledge or management experience, rather than on hard data, are often referred to as the *Seven Management Tools* (see Bakerjian

6. We won t be discussing histograms and scatter diagrams; we haven t found them well suited to most of the problems we studied.

1993, 2–11; Brassard 1989; and Mizuno 1988) to differentiate them from the Seven QC Tools. These seven tools are Affinity (KJ) Diagrams, Interrelationship Digraphs, prioritization worksheets, tree diagrams, matrix diagrams, Process Decision Program Charts, and activity network (arrow) diagrams. What perhaps is most unique about them is that they provide a method for capitalizing on experience in an orderly, methodical, fact-based way. Hence these management tools help us continue to base our choices on facts rather than the emotions to which we often revert when we lack data—or when we lack an orderly method for approaching a problem or investigation.

We've described how we've used the first of these tools. The references provide a more detailed description of how to apply all seven of these tools, together with a tips on when and how to use them most effectively.

A PROBLEM-SOLVING PROCESS

Given these problem-solving principles, and examples of tools that support these principles, it is easy to design a logical problem-solving process:

1. Clearly state the problem—in the *customer's* terms
2. Generate a list of root causes
3. Prioritize root causes
4. Develop solutions to address the largest causes
5. Prioritize solutions
6. Deploy the best solutions and measure the gains

Two Examples

We now consider some examples that demonstrate how we applied some of these tools to actual problems. The first example describes a development organization's problem with the frequency and number of new or changed designs that they released to manufacturing. This example describes how one of our software development teams used these tools in their quest to improve their development process dramatically.

Example 1—A Hardware Development Team's Problem with Design Changes

A few years ago, one of our development groups was just starting their Quality journey. They had completed their first Quality training, a course that taught the basics of Quality management and introduced several of the old data-based tools. They were raring to find an opportunity to try out their new-found skills.

Step 1: Clearly state the problem—in the customer's terms

Factory engineering was concerned about the frequency that the designers were releasing new or changed design information to manufacturing. Stated in the customer's terms, factory engineering was concerned with the number of times designers were releasing changes to the design information.

Step 2: Generate list of root causes

To examine this problem the team decided to use the root cause diagram to help them determine the reasons leading to each of the 150 design changes. The team reviewed each of the changes to determine the root cause. They gathered the changes in six categories, as shown in the root cause diagram in Figure 6.7. The largest reason for changes was what they referred to as OOPS, their euphemism for a mistake of any sort. Fifty-eight changes, more than 38 percent of the total, had been issued because of mistakes.

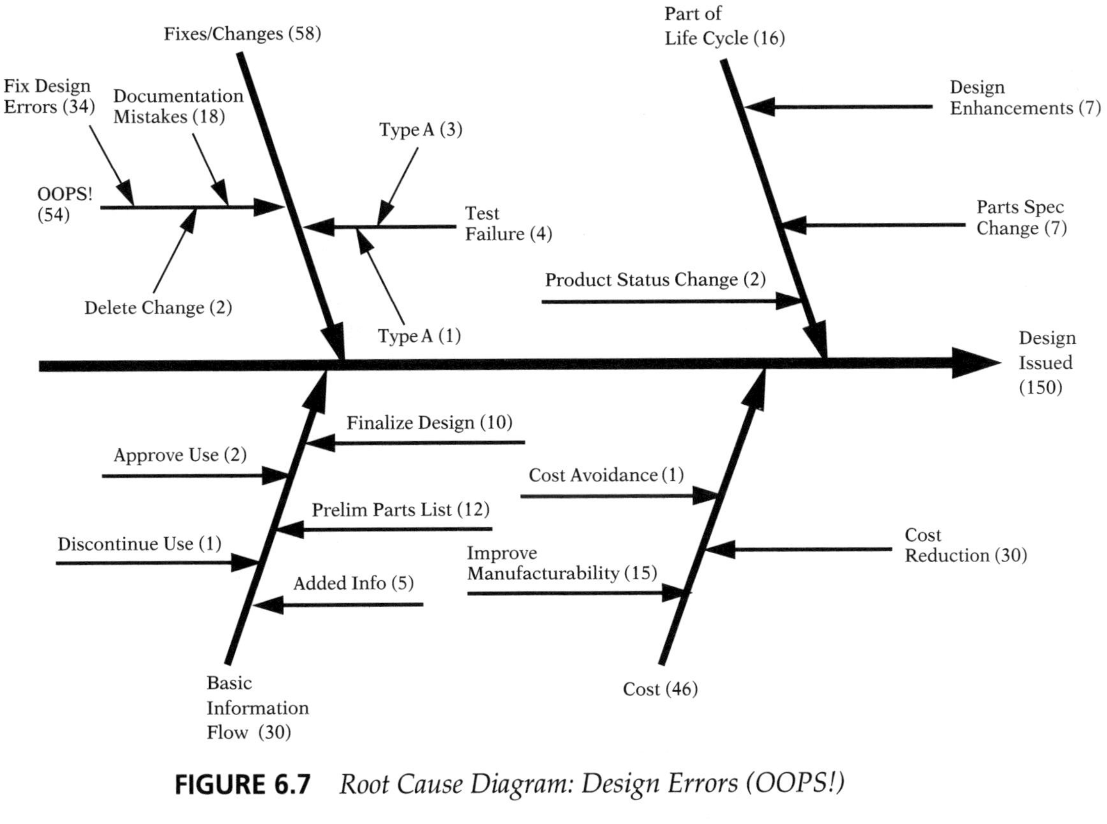

FIGURE 6.7 *Root Cause Diagram: Design Errors (OOPS!)*

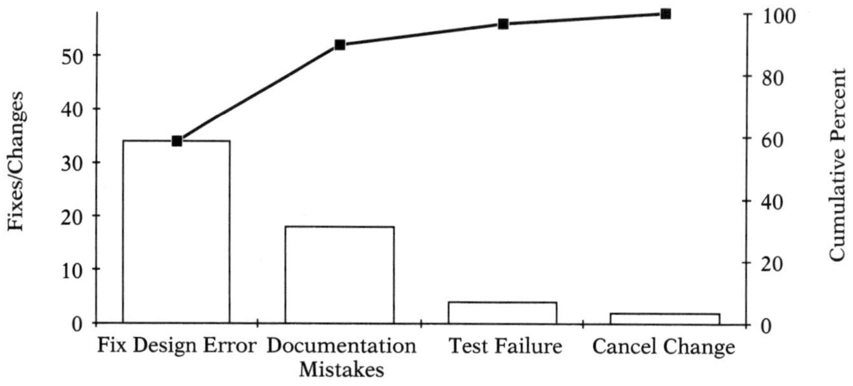

FIGURE 6.8 *OOPS! Errors: A Pareto Chart*

Step 3: Prioritize root causes

The mistakes, however, were broken down further into the categories shown on the various bones in the fishbone. These bones were then displayed in a Pareto chart, Figure 6.8, to show the causes in priority order. Clearly one of the primary causes was those listed in the OOPS category. Though not clear from the diagram, most of these mistakes were errors in the parts list—essentially typos. The problem of typos, while it was the second most frequent problem, was tackled because the team believed they could be corrected with little effort.

Steps 4 and 5: Develop solutions to address largest causes and prioritize solutions

The team then developed and prioritized solutions. They chose a two-pronged approach. The first solution was simply to focus on improved reviews of changes before they were issued. Second, they included capability in the computerized design change tool, under development at that time, to provide the related device data automatically once the part number was entered. This approach enhanced the team's ability to proofread the change; they could enter one parameter and check that the correct related information appeared.

Step 6: Deploy best solutions and measure the gains

The improvements were implemented and tracked over time. The results were displayed in a simple run chart, Figure 6.9, showing the reduction in design errors with time.

Example 2—A Development Team's Quest to Reduce Software Faults

This example is based on the experience of one of our software development teams. The team had been through the learning experience that led to their de-

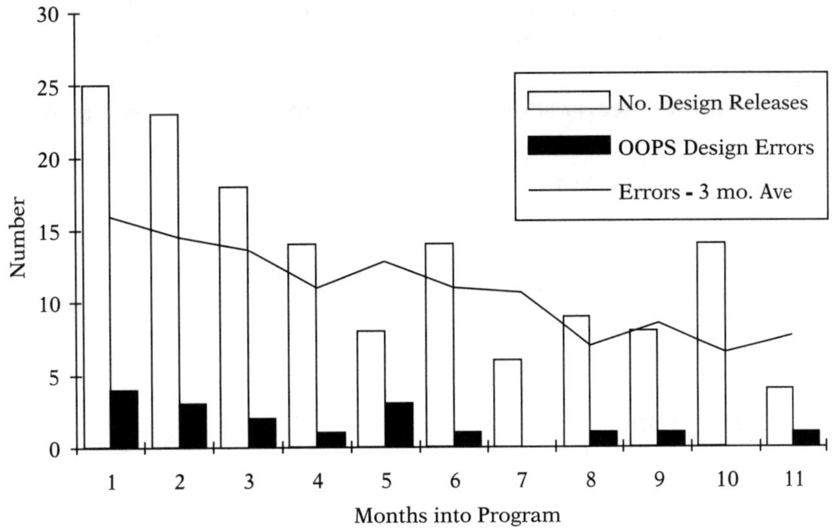

FIGURE 6.9 *Run Chart: Reduction in Design Errors with Time*

veloping, documenting, and starting to improve their process. Their high-level process, shown in Figure 6.10, starts with documenting the product definition and ends with readiness to ship the product on successful evaluation in the field.[7] While they believed that their performance was competitive, the team members were concerned that both the business and the competitive environment required them to increase both their productivity and the quality of their work dramatically.

Before they could begin, the team first had to answer the question Who is my customer? In this case, the team decided that they had two customers that they were most concerned about: the user of their software product, as represented to them through their product management, and their development team members, who were frustrated with their rate of progress in improving their productivity and quality. They decided that the primary problem was to reduce the software defect density. They also decided that, if they could suc-

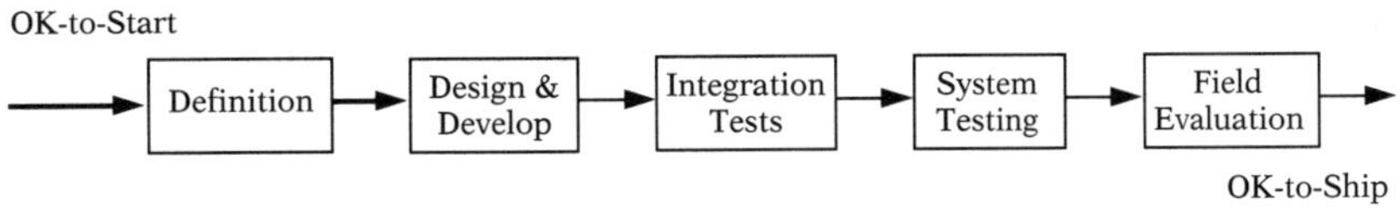

FIGURE 6.10 *Simplified High-Level Software Development Process: A Flowchart*

7. The process consisted of additional levels for each major block; we show only the highest level for simplicity, and since that is all that is required for the example.

cessfully satisfy their external customer, they would also solve their problem with their team's frustration.

Step 1: Clearly state the problem—in the customer's terms
Restating the above decision, their product management was concerned with the defect density of the software.

Step 2: Generate list of root causes
Because the team was using a well-understood development process, they decided to use the repeated questioning technique. During the analysis phase for each problem found, they religiously asked the questions, where *did* we find the problem, where *could* we have found the problem, and where *should* we have found the problem.

For example, consider the case when a feature was found not to work according to the specification during the system test phase. The answer to the first question then was, "System testing." They then addressed the second question by saying they could have found it in Integration Tests, and examined the adequacy of the test screens in that phase.

Step 3: Prioritize the root causes
The team began to find evidence that the problems were of three types: coding, design, and breakage (i.e., the fix of one problem resulted in yet another fault). The top three contributors, normalized to the largest contributor, are shown in Figure 6.11 in a Pareto format.[8]

Step 4: Develop solutions to address largest causes
The team developed a number of potential solutions, several of which are shown in Table 6.2.

TABLE 6.2 *Potential Changes to Reduce Defect Density*

- Reuse (rather than develop new) code
- Introduce process effectiveness metrics
- Introduce "lessons learned" data base
- Regularly analyze trends in fault discovery and closure rates
- Make the fault classifications more meaningful
- Use tenure development teams
- Improve designer skills
- Use design/code reviews

8. Note that this is, technically, not a Pareto chart, since it contains neither data on all the defects nor the cumulative percentage of the various parts.

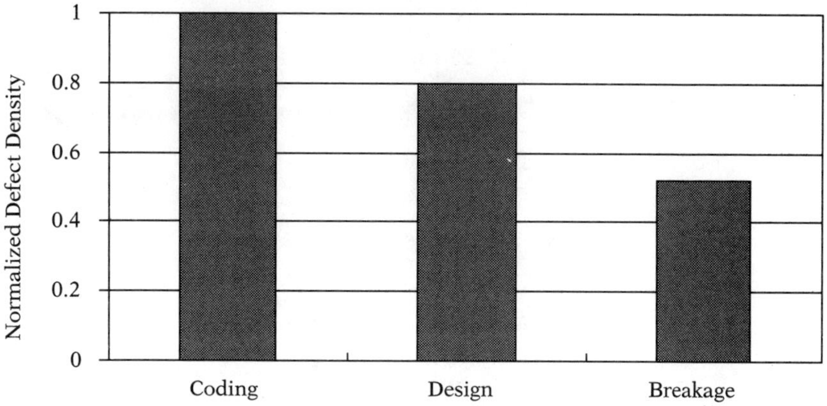

FIGURE 6.11 *Software Defect Density by Root Cause: A Pareto-like Chart*

Steps 5 and 6: Prioritize solutions and deploy best solutions and measure the gains

Over time most of the potential changes were instituted in one form or another. However, for our purposes, we want to consider the period when the team implemented code reviews as part of their design and development process. Run charts in Figures 6.12 and 6.13 show the gains associated with the introduction of code reviews. Figure 6.12 compares the fault densities in the

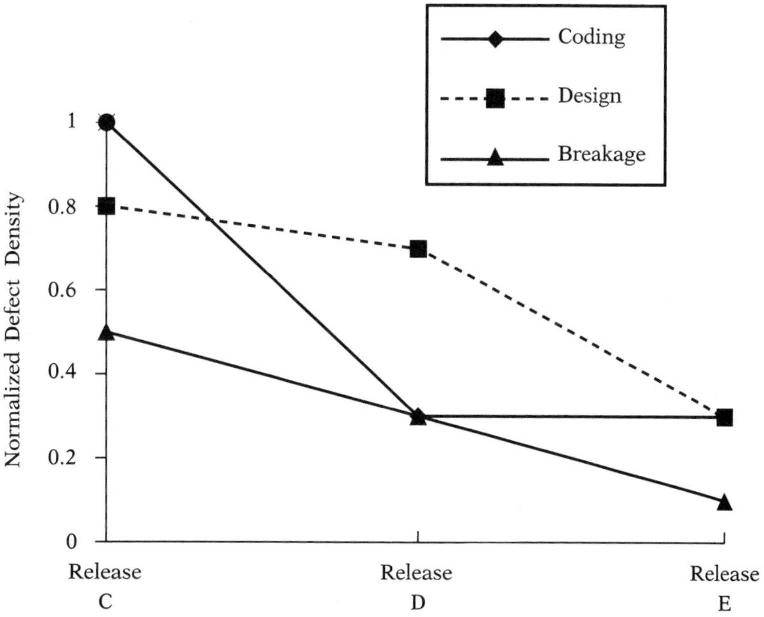

FIGURE 6.12 *Defect Density Reduction by Root Cause: A Run Chart*

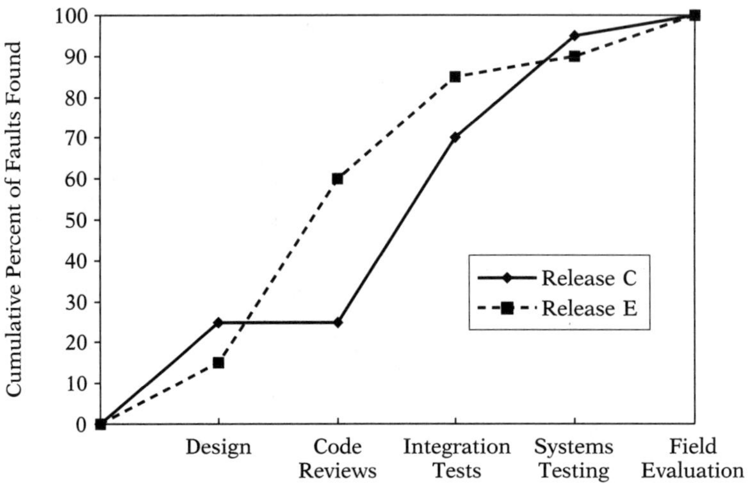

FIGURE 6.13 *Cumulative Distribution of Where Faults Found: A Run Chart*

Releases D and E subsequent to the introduction of code reviews in Release C; note that, while the review seem to continue to reduce design and breakage problems, the flattening of the curve for the coding faults, after the 70 percent reduction, implies that there are likely to be other causes associated with the remaining coding faults.

Figure 6.13 shows the impact of where, in the development process, the faults were found between Release C and Release E due to the introduction of the code reviews in the design process. Note that the faults found subsequent to the design and development process were reduced from about 75 percent to 40 percent.

Figure 6.14 uses another run chart to depict the reduction in fault density over time. In particular, the team reduced the fault density to 10 percent of what it had been at the start of a four-year period. The team began focusing on process documentation and improvement after they shipped Product A. They introduced the code review process after they completed Product C, with the results shown in Figure 6.12 for Products D and E. In addition to this order of magnitude decrease in faults, the net result of all their process efforts over the four-year interval was to reduce product development interval by 50 percent. There was also a significant increase in pride and morale of the development team.

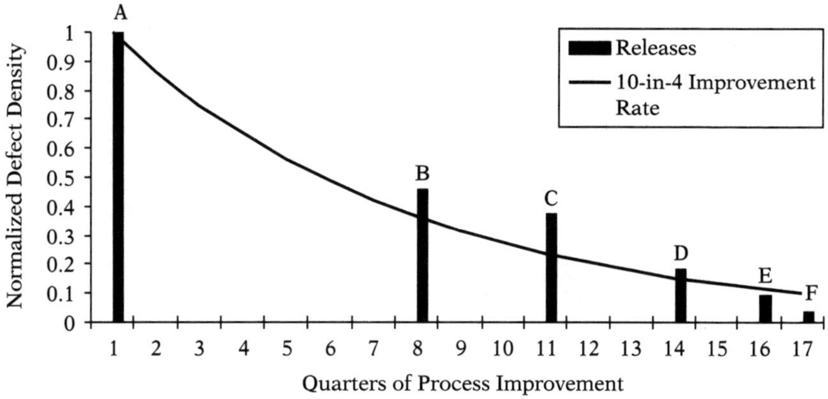

FIGURE 6.14 *Reduction in Defect Density for a Series of Releases over Time for Product Family*

7 Concurrent Engineering in a Process-Centered Environment

Strategic Quality Ideals

✓ Total Employee Involvement and Empowerment
✓ Focus on Process: the How of Doing Work

Organizational Enablers

✓ Strong Values Shared by All
✓ Reward Structure Linked to Contribution

> *Concurrent Engineering is a systematic approach to the integrated, concurrent design of products and related processes, including manufacture and support. This approach is intended to cause the developers, from the outset, to consider all elements of the product life cycle from concept through disposal, including quality, cost, schedule, and user requirements.*—Institute for Defense Analyses Report R-338

We traveled a rocky road on our journey to implement concurrent engineering. When we started, the term *concurrent engineering* was not common terminology, and no easy-to-follow instructions were available. No easy directions exist today, either. Each implementation of concurrent engineering is unique, and adds to our previous understanding. Perhaps our story will help you avoid some of the pitfalls we encountered, and shorten the time it takes you to put in place a working concurrent engineering environment.

A CE STORY

1985–They're Out There Somewhere

We worked at Bell Labs in Middletown, New Jersey. This location was the heart of product design for AT&T's small telephone systems. The factory was in Denver, Colorado. As developers, we had very little interaction with the factory during new product development. We knew the factory was out there somewhere, and for most of us that was enough.

We also had the attitude that what we didn't know, we could find out ourselves. We had control. The product would look, perform, and cost the way we designed it. We felt ownership and accountability for our products, and a lot of prestige came with the job.

Moreover, management valued designers for their individual contribution, and promoted designers who were noted for one or more significant individual achievements, or who took the lead on important projects. Designers who would step forward, take personal risks, and demonstrate individual excellence were rewarded.

The reward system (performance appraisal and salary administration) was based on a process of ranking individuals across an entire laboratory. In our case, this meant several hundred people. Managers would gather and debate, in what was a painful process, to rank those under them. Such a process fostered the value of individual excellence.

We did not overuse the term *team* back then, either. In fact, *team* was just a euphemism for *committee* and compromise.

We did work with the factory to fight fires with current production, although very few designers liked to do current engineering. Who would want to spend time helping the factory make the design easier to manufacture or cost less when you could be working on leading-edge technology, or designing a new product? Besides, there was little opportunity to be recognized for individual contribution or technical leadership if you were doing current engineering. The system just didn't recognize or reward that kind of work.

Tele-conference calls for current engineering with the factory were not boring, though. We often had emotionally charged, antagonistic exchanges. The phone has a way of exaggerating the "us" and "them" relationship, and we frequently left those tele-conference meetings feeling that the relationship was not as good as it could be.

We did have a tour of the factory once. It was impressive to see all of the machines and people putting our designs together. To see hundreds of something that once was just an idea in your mind is really exciting, and was the most memorable part of our trip out west.

We threw designs "over-the-wall" back then, and it was not considered a sin. The factory addressed manufacturing processes after we passed the design information to them. Arnold Putnam described this mode of operation as the "venerable American tradition," that is "to design first and do manufacturing engineering and quality control later" (Putnam 1985, 139). For us, it was "business as usual."

We characterize the working relationship between design and manufacturing during this period as:

- Little or no interaction during new product development
- Design owned solely by the developers; separate design and manufacturing documentation
- Individual contribution highly valued and rewarded
- Over-the-wall design transfer and serial design and manufacturing activities considered normal mode of operation

1986—You Want to Check My Work?

In 1986, the first indication of change arrived. Factory engineers caught wind of a concept called *Design for Manufacturability* or *DFM*. The stirring in the development community was that factory engineers were going to review the designs for manufacturability. A formal notice went out to all design engineers breaking the news; before any designs could be released to the factory, R&D designers had to review their work with the factory engineers at a formal DFM review.

This program did not sit well with many designers. We would often hear designers say "the nerve, to think that we would design something that you couldn't build is ridiculous," and "this is just another way for the factory to dump on our designs."

The first meetings were tense, and we were faced with more DFM issues than we had anticipated. In several cases, resolving their DFM issues would have an impact on our development schedule. The factory engineers wanted the designs reworked, and that would take time. A specific instance surrounded a printed-wiring board layout. The path spacing between the solder leads for components was wrong. Wrong in this case meant that the soldering yield would not be as high as the factory engineers wanted. Making this change would require not only engineering work but testing as well, since the board had to pass certain FCC requirements. We would miss *our* committed hand-off date to the factory.

We soon learned that the manufacturing engineers even wanted to trade off material costs of the product with assembly costs. We were not responsible for their assembly costs, but we were responsible for meeting our committed cost targets to product management. This was getting out of control. In some cases, we just ignored the manufacturing issues and sent the design the way it was.

We could always rationalize this action based on schedule or material cost objectives.

Ignoring manufacturing issues and the factory did not last long. Management sent a formal notice requiring that a manufacturing engineer had to sign the design transfer cover sheet before a design could be released to the factory. In other words, they could refuse to accept designs that they deemed unmanufacturable—if there was such a thing. We all knew that it was time to change.

The relationship between design and manufacturing during this period could be characterized by:

- After-the-fact reviews; meetings with manufacturing were conducted after the design work was complete
- Designs were frequently reworked to accommodate manufacturing issues, resulting in longer development intervals
- Relationship between manufacturing and design changed from little interaction to an antagonistic, emotionally charged and tense relationship

1987—Back to School

If there was something that we didn't know, and we needed to know it, we learned it. Designers flocked to DFM classes. Boothroyd and Dewhurst became well known, and DFM was expanded to cover DFA (Design for Assembly), DFT (Design for Test), DFE (Design for the Environment), DFS (Design for Safety, for Simplicity, and for Serviceability) and other assorted things. In fact, the list became sufficiently long that an *X* replaced the last letter as a generic variable (DFX) to include all of the downstream processes (Gatenby 1988).

DFX techniques sprang up all over the place. We were counting parts, using snaps (screws became a no-no), thinking about the orientation for assembly, making parts common between products and within products, etc., until we almost forgot that we were designers, not manufacturing experts. Some designers began to view the DFX effort as a distraction from our original focus on functionality and performance.

To make matters worse, the factory decided to help us out by writing down all of the DFM rules. That way we would know ahead of time what the DFM reviews would cover. It wasn't so bad when the first issue came out and was only several pages. Later it grew to be half an inch thick.

The working relationship between design and manufacturing during this period had the following attributes:

- Manufacturing provided DFM information via documents, i.e., DFM guidelines; reviews and rework were reduced
- Designers became more knowledgeable and sensitive to manufacturing and other downstream processes affected by the design; product designs began to reflect manufacturing issues up front in the design process, and some full-stream optimization occurred
- Manufacturing engineers remained isolated from the design process, with the exception of manufacturability reviews
- Designers began to realize that it is difficult to keep up with and handle all of the manufacturing information; awareness increased that designers cannot do DFX by themselves

1988—Part-Time is Better than No Time

It was apparent to most of us that education was not the answer. We just could not keep up. We shared the idea with manufacturing that they should participate in the design processes at key phases. This would take some of the pressure off of us, and they would have ownership for some of the decisions made about the design.

We started this out as a trial. We would invite a manufacturing engineer to New Jersey for several weeks. In one of the first pilots, the manufacturing engineer stayed most of the summer. This turned out to be a good experience for all of us. We addressed the right manufacturing issues the first time around (at least most of them that could be resolved during the short period of the visit), and folks were generally pleased.

We did have a few problems. It was a bit awkward that the manufacturing person had to up and leave home to do this. The communication systems were not adequate to support doing this type of work 1,600 miles apart. Co-location became a hot topic. Because of the cost of implemention, it never became anything more than a hot topic. Moreover, some of the more forward-looking folks among us observed that if we were ever to be successful as an international company, we should begin by learning to do "DFM at a distance" (Wesner 1991).

We also suffered through this new approach because the development process was ad hoc. All the designers did whatever they wanted, and usually selected a different approach each time. We would be hard pressed to say that we even had a process. Engineers and supervisors kept the design methodology in their heads, and if you wanted to do something new and different, you just did it. This ad hoc process certainly did not include manufacturing-related activities. The terminology we used was also different than that used in manufacturing, so misunderstandings were common.

The manufacturing person also had different priorities and goals, and manufacturing engineers were rewarded based on different criteria. They seemed very focused on cost reductions and manufacturability. There seemed to be little appreciation for technical innovation. In fact, many debates centered on the dilemma that if we just kept designing things that the factory could build, we would never design anything new.

One other small detail: the design and manufacturing tools were not designed to work together. In many cases, no matter how much we worked together, design information had to be manually retyped into a factory computer system. This was the case for plastic part drawings, stock lists for material ordering, assembly drawings that showed how the pieces went together, and other design information. Almost every piece of information had to be reworked in some way before it could be used by the factory.

We recognized that we were not there yet. Too much was missing. It was then that we put some real horsepower behind making the design and manufacturing interface work.

We would characterize the working relationship between design and manufacturing during this period with the following attributes:

- Manufacturing became involved early in the design process as an active participant (still part-time); reviews and rework eliminated
- Frequent interactions occurred between design and manufacturing; communication pathways were opened
- Concurrent design of product and the associated downstream processes took place on a limited basis, with some shared decision making between design and manufacturing; manufacturing readiness intervals (final design data to production ship) began to decrease
- Success factors for concurrent engineering began to surface, including shared objectives, communication tools, shared design tools, common values and reward systems, and common processes, etc.
- Designers continued to "own" the product design

1989 through 1992—We're in This Together

We began to turn things around in 1989 when an initiative called *Denver PRP* was started. *PRP* stands for *Product Realization Process*. The intent behind this initiative was to re-engineer the process, spanning design concept to product ship, to be faster and produce higher-quality designs. The prime targets were eliminating queue times, sharing information between design and manufacturing throughout the development process, and making activities as parallel as possible (namely, doing concurrent engineering). Nearly eighteen months were spent re-engineering the process (using a method similar to that described in Chapter 5), and another nine to twelve months were spent helping projects get on board with the new CE process and tools. We worked to incorporate all of the learning over the previous four to five years, as well as the best of what we could learn from benchmarking others. We strived for a *seamless* design and manufacturing process, a single set of documentation, a cross-functional team approach, and frequent, informal communications.

The Bottom Line

The best part of this story is that it has a good ending. Following our joint re-engineering efforts with manufacturing, we were able to demonstrate improvements across the board for printed-wiring board developments. Figure 7.1 shows the overall reduction in time-to-market measured from hardware specification complete to production ship. Figure 7.2 illustrates the reduction in the manufacturing readiness interval (time from final design information available to production ship) achieved. We reduced design iterations from

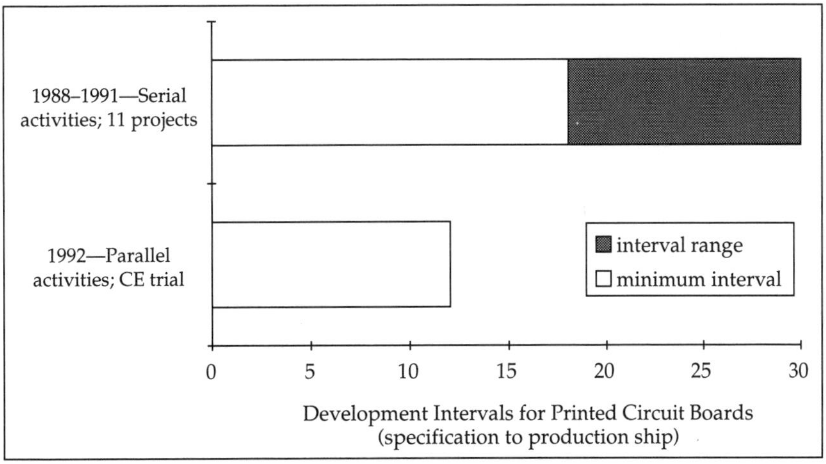

FIGURE 7.1 *Development Intervals for Printed Circuit Boards (specification to production ship)*

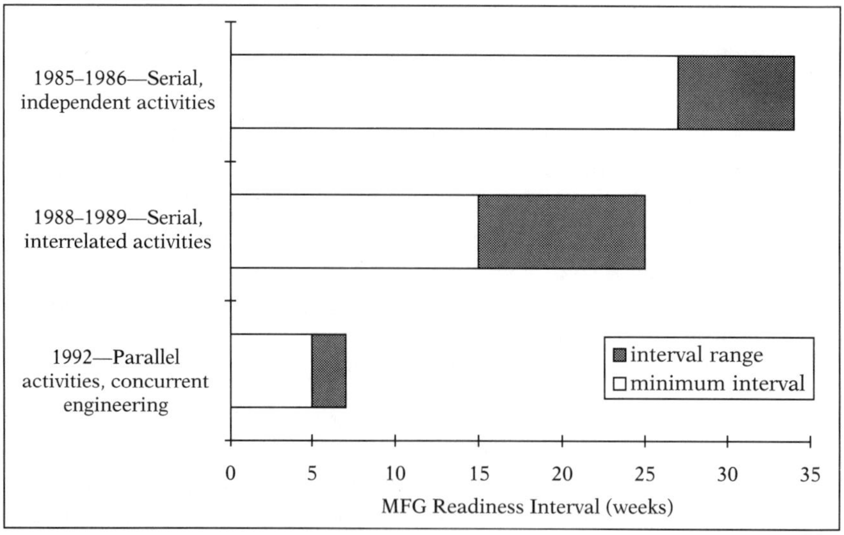

FIGURE 7.2 *Manufacturing Readiness Interval (final design data to production ship)*

three to four cycles to nearly one, at a per project savings of nearly $200K (we estimated savings at approximately $90,000 per iteration removed from the process).

When we look back over this seven-year period, it is apparent that we went through an evolution, beginning with little or no interaction between design and manufacturing, and ending with cross-functional development teams assuming end-to-end ownership for the development of both the product and the process (see Figure 7.3).

What did we learn? First, we fully realized the value of addressing not only product performance and functionality during product design, but also all of the downstream processes (as espoused throughout the DFX work). Concurrent engineering became widely accepted and supported by management as an important cornerstone to optimizing full-stream costs, reducing time-to-market, and providing quality products and services that meet user requirements (managers especially bought in after they saw the performance results from our trials as well as other published success stories).

Second, we recognized that several ingredients had to be combined to make concurrent engineering work. These ingredients are what we term the critical success factors for concurrent engineering. They are shown in Figure 7.4. The following sections detail our experiences making these success factors work.

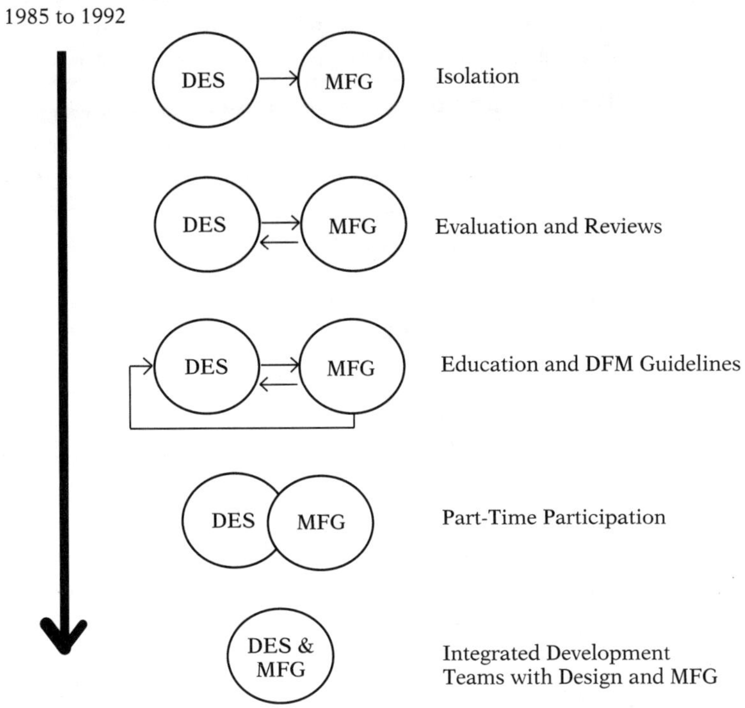

FIGURE 7.3 *Evolution toward Concurrent Engineering*

CRITICAL SUCCESS FACTORS FOR CONCURRENT ENGINEERING

The success factors for concurrent engineering in Figure 7.4 cover two areas:

- CE team structure
- CE process and tool environment

These areas are rooted in the successes and failures we experienced implementing CE. We believe that you can be successful at your concurrent engineering efforts if you pay attention to these two critical areas for concurrent engineering. We begin with people and teams first, and then discuss the role of process and tools in the CE environment.

CE Team Structure

We were a little slow to learn that concurrent engineering is more than a few designers thinking about manufacturability on their own. In fact, we could not

Critical Success Factors for Concurrent Engineering
Team Structure
• Cross-functional teams for product and process development with skilled people from Design and Manufacturing
• Open and frequent communication
• Shared values, objectives, and rewards
Common Processes and Tools
• Joint, seamless processes between Design and Manufacturing
• Common tools for sharing information

FIGURE 7.4 *Critical Success Factors for Concurrent Engineering*

begin to educate designers fast enough, and on a continuous basis, to keep up with the changes in manufacturing technology. We discovered that only a skilled, cross-functional product development team can effectively impact the design to optimize all downstream processes, and at the same time, develop the manufacturing processes to be used to produce the product. The areas that need special attention for an effective CE team include:

- Team formation and start-up
- Open and frequent communication
- Shared values, objectives, and rewards

Team Formation and Start-Up

The composition of an effective CE team should reflect those primary areas of the product's design, manufacture, and support. This team will comprise the body of knowledge used to consider all elements of a product's life cycle during the product and process development. It goes without saying that the team should be formed with people with the right skills in their functional area for the job. No process or team model can compensate for inadequate experience, knowledge, or skills.

During the formation of CE teams, you will need to be sensitive to team size. If the team should grow beyond eight to ten people, it may be useful to designate certain key players as *core team* members, and the other important contributors as *extended team* members. The roles of the core team members would be more pro-active and participative, while the roles of the extended team members may be subject matter expert or consultants. An example of a team

Sample CE Team

Team Member	Function
Electrical designer	Responsible for the electrical design and schematic for the product
Physical designer	Addresses physical circuit layouts, physical enclosures, and packaging for the product
Material management	Manages the availability and ordering of parts
Circuit pack test engineer	Develops and maintains the test processes and equipment for the product
Manufacturing or product engineer	Manages the introduction of the product into the factory, and conducts on-going change management for the product
Manufacturing process engineer	Develops and maintains the assembly processes for the product
Purchasing buyer	Works with and procures from outside vendors subassemblies or fabricated parts
Servicing support engineer *	Provides support to customers and field technicians when the product needs repair or replacement

* Extended team member

FIGURE 7.5 *Sample CE team for printed circuit board development*

we formed for printed circuit board development is shown in Figure 7.5, with the extended team members or consultants (subject-matter experts) denoted by an asterisk.

We struggled with the concept of core team and extended team members because we wanted everyone who had direct influence on the success of the project to be (and feel that) they were part of the team. We wanted them to share in the objectives for the team as well. As a result, many team members who only worked on the project part-time (i.e., were not dedicated to this one project) were also considered core team members.

With one of the CE projects we supported, almost 75 percent of the team members were part-time. They came from areas like purchasing, material planning, servicing, etc. This pilot team was uncertain what role the part-time members would play. The questions from full-time members were "Should the part-timers come to every meeting?" and "Do we really need to list them as team members?"

In one case, the full-time members questioned why a person from material planning should need to be at every weekly meeting. The argument usually went something like this:

> *Why should the material planner attend every meeting during the very early design of the product, when we don't even have a list of the components we want in the product? They are very busy, and this seems like a waste of time for them. They just sit there anyway and don't say anything.*

Interestingly enough, the material planner wasn't posing this argument. The engineers were making assumptions for them. When asked directly, the material planner responded by saying:

> *I realize that I don't contribute actively at every meeting. However, for the first time I am involved early enough in the process to provide input to components as they are added to the design, rather than wait for the design and parts list to be complete (and nearly impossible to change). I really want to be at the meetings.*

When the team heard this response, they were happy to have the material planner attend every meeting. In fact, as this process played out for each team member, it helped the team sort through the roles for each person, and clearly established the value that each team member was bringing to the table.

Start-Up

When a new project is initiated, the members of the CE team should come together at a start-up meeting. All involved parties (full- and part-time) should attend. The agenda should include:

1. Introduction of the project to all team members
2. Review of the project concept and business objectives
3. Detailed review of the CE process to be followed
4. Discussion of how each team member would like to be involved during the project
5. Communication mechanisms for the team

Agenda Item 3, the detailed review of the CE process, is essential before the discussion of roles and involvement (Item 4) is held. Once each person understands what will happen, they should be able to state their preferred mode of interacting with the project. Some full-time members will be involved in every meeting. Part-time members may prefer to be heavily involved during specific phases, and not involved at all during other phases of the project.

Open and Frequent Communication

Once the right team members are brought together, and they have worked through the roles of full-time vs. part-time members, they need to agree on how they will communicate with one another throughout the project. Effective communication cannot be overstressed for CE teams. During the start-up meeting, the team should devote a significant amount of time to this topic, and should review all the communication means at their disposal, including:

- Frequent meetings
- Electronic mail
- Video teleconferences
- Voice mail
- Faxes
- Paper mail (and overnight / express mail)
- Telephone calls
- Shared computer files (networks)

The team should agree on what methods will be used for each type of information exchange they can identify, including: document distribution, meeting notices, meeting notes, requests for assistance between team members, product design information, and so on. The team should brainstorm the different scenarios and pick the preferred mode of communication for each.

Once the mechanisms for communicating have been selected, it is equally important to ensure that every team member has equal access and knowledge for using the selected communication tools.

One CE team we supported selected voice mail and e-mail as the primary communication methods. One half of this team was in New Jersey, and the other half in Colorado. The engineers in Colorado liked voice mail, and were not accustomed to electronic mail. The engineers in New Jersey, on the other hand, primarily used e-mail, and voice mail was used less frequently.

This team faced two issues with their communication choices. First, engineers accustomed to one form of communication had difficulty switching to the other, and this resulted in a responsiveness issue. They simply were not used to checking messages on both systems. People used to frequently checking e-mail continued to check e-mail frequently, and the same applied to voice mail users. The result was that engineers who sent voice mail to e-mail users did not receive quick responses, and vice versa. It took the team several months at the beginning of the project to sort this out; much too long given the importance of this early design phase.

Second, both groups lacked a mutually acceptable protocol for communication on a day-to-day basis. Upon receiving an e-mail or voice mail, some people would work the issue before responding, and when they finished solving the problem, they would get back to the other person (sometimes days later). Since the sender did not hear back right away, they would seek help elsewhere, not knowing the problem was being addressed. They would also develop the impression that the first person they requested help from was unresponsive. We addressed this problem by formally documenting and socializing a communication protocol (see Figure 7.6).

We followed up on this protocol with a measure of responsiveness in the form of a survey and monthly report card. Each location would evaluate the other location on a scale of 1 to 5 (1 being poor, 5 being excellent). In addition to responsiveness measures, the monthly report card also reflected performance in several other key areas: percentage of commitments met, and number of design changes to fix design errors. A sample survey and report card are shown in Figures 7.7 and 7.8.

We tracked these data every month, and developed a composite score based on the data. Figure 7.9 shows the performance improvement in this composite

Communication Protocol

If you receive a voice or e-mail message, please

—Let the sender know that you *received* the message as soon as possible, and
—answer their question in your response, OR
—tell them that you will work the issue and get back to them by a certain date, OR
—tell them that you are not the right person to work this issue.

It is not acceptable to NOT respond, even when you are not the right person to work the issue.

If you are the sender of a voice or e-mail message, please

State exactly what you need in your message, and when you need it (if possible, avoid just saying things on voice mail like "call me back"). Let them know why you are calling and what you want. Be sure to tell the person how to get in touch with you (phone number or e-mail address, and times they can call you).

FIGURE 7.6 *Communication Protocol for e-mail and voice mail*

score over time. Many of the improvements stemmed directly from quality improvement teams working in specific areas. Some of the improvement is believed to be driven by the very nature of collecting and socializing the data. The data collection effort shows the community what management values, and this value statement eventually drives behavior.

Bi-monthly Survey on Responsiveness and Communication

Based on your day-to-day interactions, rate your counterparts on the following areas using the scale 1–5 (1=Never, 3=Sometimes, 5=Always)

1. When I leave a message via voice mail or e-mail my counterparts are quick to respond to me. (1 2 3 4 5)
2. When I request assistance or evaluation of an item, my counterparts are efficient and helpful. (1 2 3 4 5)
3. My counterparts are willing to make commitments. (1 2 3 4 5)
4. My counterparts follow-up and close out items. (1 2 3 4 5)
5. When I am working with my counterparts there is a professional atmosphere and they are easy to work with. (1 2 3 4 5)

I think my counterparts responsiveness and communication skills are:
__ improving __ staying the same __ getting worse
I would like to nominate ________ for a teamwork recognition award.
Reason:

FIGURE 7.7 *Responsiveness Survey*

Responsiveness and Communication Report Card	Grade for the Month of April
Communication effectiveness	C+
Responsiveness to requests for assistance	B–
Willingness to make commitments	B–
Ability to follow-up and close items	C+
Ease of working with	A–
Commitments made (B = 70–80%)	B
OOPs EDIs—design changes due to defects (B = 2 per month)	B

FIGURE 7.8 *Responsiveness Report Card*

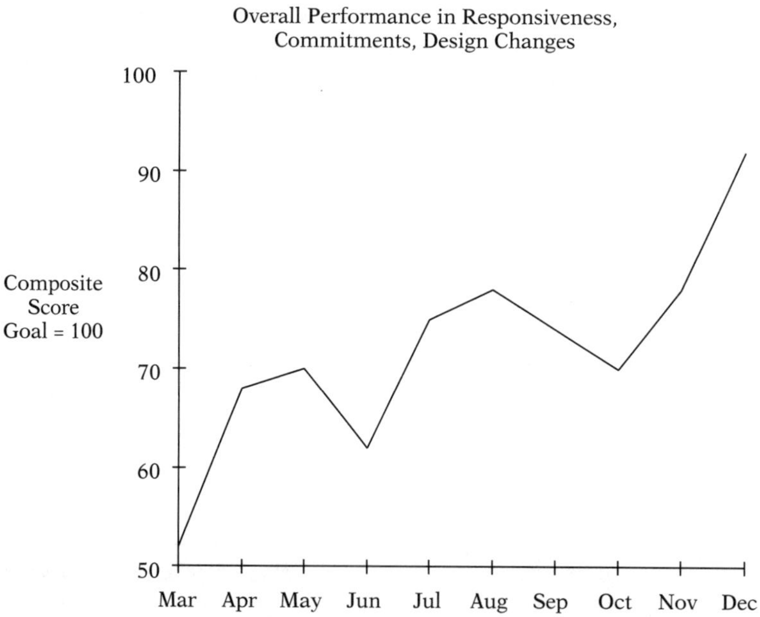

FIGURE 7.9 *Performance Tracking of Responsiveness and Other Key Report Card Metrics*

We also added to the list of communication options the virtual disk drive for personal computers. Virtual or shared drives are now common on most PC networks. A *virtual drive* is a shared drive that resides on a network machine, sometimes called a *file server.* Any person on the network can have access to files placed in specially designated shared directories, thereby allowing people to work together on documents. In our case, a special gateway and dedicated telephone line were established between New Jersey and Colorado that allowed this same type of file sharing to occur 1,600 miles apart.

Meetings

The number and frequency of meetings goes up with a CE team. This is viewed as negative by some. However, we have observed that the most effective CE teams work together closely and meet often. For our pilot program we suggested that the entire team meet at least once a week, and encouraged them to meet more frequently (every day) during key phases of the project. Some members on this pilot program responded badly to this meeting arrangement. A commonly heard complaint was "I am too busy to meet that often." As the team became actively involved in the CE process, and saw the need for continuous and frequent communication and negotiation, they rapidly gained an appreciation for meeting frequently. Rather than complaining about meetings, they focused their efforts on making the meetings short but effective. The meeting format that worked well for this team was that during the first five minutes of each weekly meeting, they would cover those issues that affected the part-time members. If no issues were surfaced in the first five minutes, these members were free to go. They also debriefed at the end of the meetings to assess what worked well with the meeting and what didn't, and adjusted the next meeting accordingly.

We also encouraged at least one face-to-face meeting during the beginning of the project. This face-to-face meeting is a good opportunity to conduct the start-up session discussed previously. This meeting also helps the team members get to know one another better, and helps them "put a face to the voice." This time spent building relationships turned out to be very worthwhile later in the project when difficult and emotionally charged issues arose.

Shared Values, Objectives, and Rewards

CE teams will be most effective when they operate under a common system of values, share objectives, and are rewarded equally for the achievement of those objectives.

R&D and manufacturing have traditionally operated in different cultural environments, and have placed value on different types of activities. For example, our manufacturing organization has always placed a high value on cost-reduction activities, whereas R&D most values the development of new and innovative technology. Manufacturing also places a high value on manufacturability and re-use of components in new products, whereas R&D has heavily focused on the performance and functionality of the new design. Gomory writes in the Harvard Business Review, "In the United States, a kind of caste system has emerged in product development. Design engineers are focused on product features and performance, which have more prestige in the engineering world. Manufacturing people are mired in the grim and gritty details of production—in intermediate costs, in the ways components are actually put together" (Gomory 1989, 102).

Individual values also play a role. For example, the following statements are defeating to concurrent engineering.

- Design engineers are *better* than manufacturing engineers
- *Individual* performance is more important than team performance

On the other hand, values that support concurrent engineering include:

- Engineering skill diversity is *recognized and valued* between design and manufacturing
- Teamwork and team skills are *as important* as individual performance

These differences in organizational and individual values have a direct impact on the effectiveness of a CE process, and they play out in the different objectives and reward systems in place for each organization. The obvious differences in objectives are those associated with the differences in values. For manufacturing these include objectives such as *percent cost reduction per year* or *percent reduction in labor and load* associated with the product assembly, and for R&D the objectives include items like *time-to-market* for new product technology, or *number of patents* received.

These differences in objectives show up during the development of a new product in the behavior of the team members and their managers. One specific example surfaced with a CE team we coached. The problem was that the team was unable to resolve issues cleanly and quickly. The scenario would play out with one team member initially taking a strong stand on one side of an issue. When that team member was unsuccessful convincing the others that their approach was the right one, they would go back and engage their supervisor in the problem.

That supervisor would then call the supervisors of the other team members, and defend their person's position. These supervisors would eventually reach a compromise decision, and inform the team. When the team discovered what was happening, and learned that supervisors who were not on the team would be making decisions for the team, they felt less empowered, and the interpersonal relationships within the team degraded.

This problem was rooted in the lack of shared objectives, and, more specifically, the knowledge that the team members had regarding performance review. They knew that when their performance evauation came around, they would be evaluated based on the same values and objectives on which they had always been evaluated. This meant that within their CE team, they needed to promote and defend those values at all cost. Since the team was evenly split between R&D and manufacturing, issues that traded off the traditional objectives of these two communities were resolved at the expense of the team's well-being and effectiveness.

To correct this situation, we worked between the team members and managers to develop a value statement and guidelines for working with teams. We had the value statement and guidelines printed on large flat magnets and distributed to all of the CE teams and managers. Figure 7.10 and 7.11 show the value statement and guidelines.

Next, we encouraged the team to adopt a single set of objectives that they would share based on these values. For example, our development organization had as an objective the design transfer date to manufacturing (called the

Performance Values

The right design and manufacturing people put together in a productive team environment will make decisions that benefit the business as a whole. Our ability to perform as a team in managing change and introducing new designs is critical to our success in achieving the goals of:

- Quality—of process and of product
- Interval—need identified to product shipment
- Cost—of product and of doing work
- Service—to our customers

Teams will perform according to how they are evaluated. The performance of our teams in these areas and the quality of their teamwork shall be the predominant measures of our success in managing change and introducing new designs, and will be the guiding values in all evaluations. Outstanding examples of teamwork will be recognized.

FIGURE 7.10 *Performance Values*

Performance Guidelines

- All activities shall be executed by a team consisting of both design and manufacturing engineers. The team will at least consist of those disciplines necessary to approve the change per our previously agreed upon EDI process (design transfer process).
- Supervisors will work to guide the team, and not necessarily to defend their team members' positions. Guidance to the team or team members should be given only if the supervisors agree, and the information is given to the team as a group.
- Supervisors will lead with principles and values instead of managing with direct decision making on specific issues.
- Supervisors, when posed with a problem, will first ask who the team members are and how the team is working together, instead of trying to understand the specific problem and developing a solution.

FIGURE 7.11 *Performance Guidelines*

Final EDI date). This was a major milestone, and R&D worked very hard to meet this commitment. On the other hand, manufacturing had as their commitment the production ship date. Before CE teams were employed, these objectives did not work well together. The R&D community would be unwilling to work manufacturing concerns before their EDI date, for fear of missing this commitment. This unwillingness to work with manufacturing early in the process required manufacturing to do all of their work between the EDI date and the production ship date. To ensure that they met the production ship date, manufacturing would often press the designers to have the earliest possible Final EDI date. So, rather than working together, both parties would push back and forth on this artificial design transfer date.

When the team members agreed to share the production ship date as the team's objective, the artificial importance placed on the interim milestone for the Final EDI was subjugated to the ultimate objective of shortening the overall interval for development, including the introduction into manufacturing.

The final ingredient needed to make shared values and objectives work is that the individuals on the team need to know that their performance evaluation and reward will be based on these shared values and objectives. If each individual believes that he or she will be evaluated and rewarded based on these shared objectives, each will act accordingly. On the other hand, if team members believe that their evaluation hinges on the "traditional" measures for their organization, they will continue to operate as if no shared objectives existed.

A performance review process we recently adopted aligns well with the concept of shared values and objectives. The performance system is based on two elements: the ability to live the values, and the ability to meet objectives. The values for this system were developed for the business as a whole. They include:

- Respect for individuals
- Dedication to helping customers
- Highest standards of integrity
- Innovation
- Teamwork

The feedback on individuals' abilities to live the values was based on input from other team members, customers, suppliers, and managers in what is termed 360° feedback (feedback from superiors, subordinates, customers, and suppliers could be graphically depicted as coming from all directions, hence the term 360° feedback). The manager for each associate collects and compiles the 360° feedback data and assesses the associates' abilities to live the values in all of their interactions. This feedback comprises half of the review.

The "objectives" element of the performance review is a negotiation process between each individual and their manager. The individual prepares a set of objectives (for full-time CE team members, these objectives should align with, or be the same as, those agreed to by the team) and presents them to their direct manager.

The manager and associate work through the objectives to make them specific, measurable, actionable, realistic, and timely (SMART), and individuals leave the meeting knowing exactly how they will be evaluated. At regular intervals throughout the year they get back together to revisit these objectives and update them, as well as to assess performance toward meeting the objectives previously set.

One hardware organization created a common set of metrics from their objectives for development and manufacturing, and thereby created a shared measurement system for both R&D and manufacturing. Figure 7.12 shows a blank table like the one tracked by these organizations.

Once you have formed your cross-functional CE team with people having the right skills, worked the start-up issues including communication, and ensured that the values and objectives agreed to by the team align with each individual's performance measures, then you have the necessary drivers for a successful CE team. The next step is to give the team an effective CE process and tool set to apply to their project.

CE Process

A CE process must achieve two objectives:

1. Design, develop, and verify a product that optimizes all elements of the product life cycle from concept through disposal, including quality, cost, schedule, and user requirements
2. Concurrently design, develop, and verify the manufacturing processes that will support the production of the product

These objectives shift the manufacturing readiness work from the tail end of the development process to be concurrent with or parallel to the development process, as shown in Figure 7.13.

SHARED METRICS	Baseline	Goal	External Benchmark	Current Performance	Trend
Defect per Unit (DPU) Metrics					
• Number of modification requests during development interval					
• Number of design iterations					
• Number of changes after production ship from design errors					
• Final test yield					
• Field return rate					
Cycle Time per Unit (CPU) Metrics					
• Interval from spec. to production ship					
• Interval to execute a change					
Cost Metrics					
• Cost of material, labor, and load (actual vs. target)					

FIGURE 7.12 *Sample Table of Shared Objectives between R&D and MFG Organizations*

In practice, concurrent engineering is not as simple as pictured in Figure 7.13, and in this section we address the details of a CE process. A CE process operates on two important principles: incremental information exchange and risk management of design data.

Incremental Information Exchange

An important process concept for concurrent engineering is the incremental flow of information. In traditional development operations, information is released in batch mode to manufacturing. The traditional trigger for the release of information is a fixed transfer date for design data. This date is often computed by calculating back from the desired ship date to determine the design transfer date. This interval, from final design transfer to production shipments, we called the *manufacturing readiness interval*. This interval contained all of the activities necessary to take final design data and get ready for production of the product, including activities such as test development, manu-

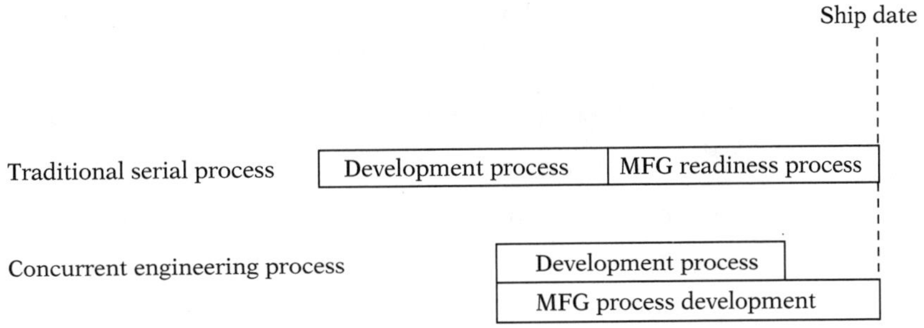

FIGURE 7.13 *Shift from Serial to Parallel Processes with Concurrent Engineering*

facturing process design and development, material structuring and ordering, packaging development, vendor selection, and so on.

Before the introduction of CE, developers assumed that the design information transferred on this date was final, and considered this to be a significant end-point for development activities. Developers staged all of their activities based on this date, and attempted to do whatever was necessary to meet it. An important objective for R&D, traditionally speaking, was the commitment date for final design information transfer (recall the discussion under team objectives).

In a CE process, information is transferred to manufacturing in an incremental approach. Design data are shared with manufacturing openly and continuously throughout the design process. The actions taken by manufacturing depend on the confidence level in the data and the type of information being shared.

For example, consider the design of an expansion card (printed circuit board) for a computer. During the concept phase of a CE process, the information that would be made available to manufacturing would include the product's functionality, purpose, size, estimated cost, and architecture. Manufacturing engineers could then begin to consider the

- Selection of manufacturing and component technology (through-hole, surface mount, tape automated bonding (TAB), leaded grid array, etc.)
- Selection of test technology (built-in self test (BIST), functional test, device-level testing, etc.)

These decisions will drive manufacturing planning and capital equipment allocations early in the process, so that requirements for new manufacturing equipment and procurement intervals will not delay the process later on.

During the design phase, the information available includes early stock lists, schematic drawings, and circuit board layout drawings. At this stage, manufacturing would be involved with:

- Component selection
- Printed circuit board test strategies
- Spacing and layout requirements for the printed circuit board

Manufacturing involvement in component selection early in the process allows the CE team to consider re-use of existing, functionally equivalent components where possible. It also allows the CE team to evaluate the performance of parts already in use on other products made in the factory to check for supplier reliability and part reliability. In addition, the material planner can forecast very early in the process what parts have long lead times, and can begin early ordering of critical components so that they do not gate the process. At the same time the material planner can begin to build the bill of materials for the board, and the purchasing agent can begin negotiations with suppliers. Involving suppliers at this early stage allows the CE team to understand the suppliers' capabilities better and allows suppliers to provide inputs into the design process. This early supplier involvement can help take material or subassembly procurement out of the critical path, and helps prevent rework later in the process to accommodate suppliers.

During the schematic capture process, the test engineer can work with the electrical designer to incorporate features into the design to optimize the production testing process. This would include items such as test points.

Given the layout drawings, the manufacturing process engineer on the CE team can help to evaluate the necessary component spacing and requirements for open space around the perimeter of the board. This will allow the CE team to optimize the product for the existing capabilities of the manufacturing equipment, or allow the process engineer sufficient time to develop new equipment fixtures to accommodate the needed component spacing.

Consider another example of the development of the housing for a computer. During the concept phase the information that is available includes conceptual drawings of the assembly and piece parts. Both manufacturing and service engineers can use this information to help evaluate the manufacturability (how will this be assembled) and serviceability (how will this be dis-assembled and repaired) of this conceptual design. In addition, they can help work through material selection issues (e.g., metal or plastic) and the relative impact this decision has on cost, assembly, repair, re-use, and disposal. At this early phase the supplier of the parts can be selected and involved in the design process (much like what was described at the beginning of Chapter 5 in the example about the plastics molder).

During the design phase for the housing, detailed piece part drawings are produced, and in some cases models are constructed. Again the manufacturing and service engineer and the supplier can address manufacturability and serviceability issues. Also at this point the manufacturing engineer can begin to design and develop manufacturing processes that will support the manual or automatic assembly of the piece parts.

From these examples you can see evidence of incrementally shared information with manufacturing and other downstream functions that allow these functions to contribute to the design of the product, and begin developing associated processes to support production or servicing of the product.

Risk Management with CE

At each stage of the design process, information can be made available to manufacturing or servicing organizations, but that does not mean they will initiate activities based on these data. In some cases, we have observed that manufacturing locations are willing to review the information and contribute to design decisions (one element of a CE process), but they are unwilling to commit their own resources to develop manufacturing or services processes in support of the product (the second element of CE).

The reason is simple. Any work they do based on early design data is subject to being reworked, or potentially being scrapped. Such has always been the case when someone chooses to work from incomplete or unstable data. So what is different with a CE process? A CE process acknowledges the risk up front and manages this risk.

Recall that one of the benefits of concurrent engineering is reduction in time-to-market. Risk management for a CE process is balancing the needed reductions in time-to-market against the risk of design changes that result in rework or scrap. Risk management for a CE process is the awareness that design data stability is not a black and white issue but rather a fuzzy gray, and the task of sorting through the fuzzy gray to establish what most likely will stay the same, and what won't, is the challenge for the CE team.

For example, consider the printed circuit board example discussed previously. During the design phase, early stock lists are available for the product. The risk management associated with this stock list considers:

- Lead time of components compared with earliest need date
- Confidence level of component (high or low), usually assigned by designer
- Quantity required for first build and calculation of dollar loss associated with long lead time parts that are ordered and then scrapped
- Impact on the development interval if long lead time components are not ordered

Equipped with these data, the CE team can now make intelligent trade offs and manage the risks associated with early design data. They can lessen the risk and extend the interval, or increase the risk and shorten the interval. In some cases, the interval can be shortened with little risk. For example, if the longest lead time component on a printed circuit board is for the most stable part, ordering this part has few risks and prevents the process from being gated by the availability of this one component.

The process of risk management becomes more sophisticated when complex manufacturing processes are developed concurrently with the product design, but the concept remains the same. The manufacturing engineers must work with the designers to assess the stability of the information, weigh the impact and risks of acting on the data, and initiate those activities that warrant the risk.

Moreover, information flow is not one way. Manufacturing process design, when done concurrently with the product design, influences the design of the product. This goes back to the first objective of a CE process: consider all elements of the product life cycle from concept through disposal, including quality, cost, schedule, and user requirements. More specifically in this case, the objective is to influence the design in such a way that production processes are optimized and the total full-stream cost is as low as possible, including material costs, labor and load costs, repair costs, and disposal costs.

The more sophisticated the manufacturing process, the more resources on the manufacturing side will be required to manage the design data and develop manufacturing processes. This helps explain the observation that many electronic products with automated assembly processes (like VCRs or Camcorders) developed in Japan using CE techniques require more process engineers than product designers during the development period (almost the inverse of many American companies).

Designing a CE Process

The method of designing a CE process is very straightforward, and given the right people assigned to the task, can be done quickly. Tool support, on the other hand, can be extremely complicated and time consuming. We recommend that a re-engineering approach be followed to create a CE process, as described in Chapter 5.

Recall from Chapter 5 that the key steps of process re-engineering include:

1. Concept development and feasibility analysis
2. Planning and definition
3. Development
4. Verification
5. Deployment

What will be discussed here are the steps to be followed by a re-engineering team to complete just the process definition element of the re-engineering process (Step 2 in Chapter 5).

Given the two objectives of a CE process as stated previously, namely to:

1. Design, develop, and verify a product that optimizes all elements of the product life cycle from concept through disposal, including quality, cost, schedule, and user requirements
2. Concurrently design, develop, and verify the manufacturing processes that will support the production of the product

the re-engineering team should design a process that meets these needs. We suggest an exercise to help the team think through the process in light of these two objectives.

Begin by writing down the high-level steps of the process. This would be a functional level description, not the detailed CE process steps. For example, the high-level steps for printed circuit board development are:

- Hardware specification
- Schematic capture

	Concept Development	Schematic Capture	Printed Wiring Board Layout
Information available	• architecture • cost target • approximate size • functionality	• early parts lists • schematic design	• board dimensions • component layout
Full-stream considerations	• component technology • assembly technology • test technology • servicing strategy • repair strategy	• component re-use • component reliability • supplier relationships • test strategies	• component spacing • board aspect ratio • spacing requirement for perimeter of board • component orientation • component accessibility for repair
Activities initiated	• capital equipment allocations and procurement • advance manufacturing planning for space, labor, etc.	• test design • component lead time evaluation • bill of material structuring and part ordering • supplier negotiations	• manufacturing process design and development • test fixture design and procurement

FIGURE 7.14 *Exercise for Supporting CE Process Design: Example for a Printed-Wiring Board*

- Printed-wiring board layout
- First build
- Integration and test

Then discuss the information that is available at each step of the process. The team should ask two questions for each piece of data or information: 1) what manufacturing, services, or other downstream operations should be considered that may impact the product design (full stream considerations)? and 2) what activities might be initiated by a downstream function that could be executed concurrently with the design of the product? The team should create a table like the example shown in Figure 7.14 for a printed-wiring board. To do this right, the re-engineering team itself would have to be cross-functional. That way the team can really sort out questions like "What would I, as a manufacturing product engineer, want to contribute to the design at each phase, and what would I want to initiate in manufacturing to begin preparing for production?"

After completing this exercise in detail for their specific process, the re-engineering team would integrate the *full-stream considerations* and *activities initiated* into the new process using a variety of methods, including:

- Document all new process steps to include "activities initiated"
- Add the additional inputs and outputs associated with these activities to each process phase
- Add full-stream considerations to entry and exit checklists or gate criteria
- Include key elements of full-stream considerations in design review guidelines

The *activities initiated* steps may be documented as a parallel process flow to the design process, where each activity is described with a purpose statement, inputs and outputs, suppliers and customers, and set of process steps as described in Chapter 4 on process management. These process flows would be shown concurrently with, and integrated into, the development process.

The result of the CE design work is a process that is common between design and manufacturing, and seamless in that no discrete design-to-manufacturing hand-off points are evident; specifically, design and manufacturing continuously interact throughout the entire product design and manufacturing process development.

At this point in the re-engineering process the team can begin the very important work of designing the tool environment (systems, software applications, etc.) for the CE process, and writing requirements for these tools. Tools will play a critical role in the success of the new CE environment. Traditional tools between our development and manufacturing groups actually made sharing information difficult, thus discouraging it.

We experienced many problems with existing tools, including multiple databases for the same information, different CAD tools in design and manufacturing, and incompatible file formats. These tool differences did not support incremental information exchange, and often forced us to reenter the same design and manufacturing data into multiple tools: essentially to perform the same function twice. Moreover, we were then forced to keep the different tools synchronized with the right data during the development process. This effort alone was difficult and time consuming.

The re-engineering team will very likely spend a significant amount of their design effort on the definition and integration of tools that support incremental sharing of design data, and the concurrent development of manufacturing and other downstream processes. In two of our re-engineering efforts the existing tools were replaced with tools designed to support a CE environment. In a third case an umbrella tool was put in place to manage the data in all of the other tools. Both of these approaches to tools resulted in the majority of the re-engineering resources being spent on tools.

Once the definition work for the new CE process and tools is complete, the re-engineering team should begin the process re-engineering development phase as described in Chapter 5. The re-engineering team can take the high-level assessment provided in Figure 7.15 to see if their process definition and planning support the primary success factors for concurrent engineering.

Quick Assessment of Concurrent Engineering Success Factors	
Rate on a scale of 1 to 5 (1 = not addressed, 5 = fully supported)	
Common Processes and Tools	
Joint, seamless processes between Design and Manufacturing	(1 2 3 4 5)
Common tools for sharing information	(1 2 3 4 5)
Team Structure	
Cross-functional teams for product and process development	(1 2 3 4 5)
Skilled people from Design and Manufacturing	(1 2 3 4 5)
Open and frequent communication	(1 2 3 4 5)
Shared values, objectives and rewards	(1 2 3 4 5)

FIGURE 7.15 *Concurrent Engineering Success Factors—Quick Assessment*

8 The Role of Teams

Strategic Quality Ideals

✓ Total Employee Involvement and Empowerment
✓ Focus on Process: the How of Doing Work

Organizational Enablers

✓ Leadership and Commitment at the Top
✓ Support and Alignment among Middle Managers
✓ High Morale and Satisfaction of Employees
✓ Strong Values Shared by All

In the preceding chapters, we have shared with you a number of concepts, principles, methods, and tools that you can apply to improve your quality and manage and improve your processes. At this point you may ask, how can I use these most effectively? In this chapter we talk about "people tools" to do just that: *teams* of various types.

TEAMS

The idea of using teams to carry out engineering development projects is not in itself new. What is new is a growing belief that teams can help improve development productivity and product quality. Three newly important foci are

- The growing acceptance of teams as the best way to organize to develop products
- An expanded view of who should be on a product development team
- The changing roles of team members *and* their managers

One of the authors recalls an experience from several years ago that illustrates how people until recently viewed teams vs. functional organizations.

Example. Our organization got a new boss. We were a department of twenty or so mechanical engineers and technicians, whose role was to provide the mechanical design support for the electrical engineers who were designing various telephones for the Bell System. As "physical design" engineers who turned their elegant breadboard circuit designs into manufacturable products, we had a pretty good working relationship with our electrical engineer colleagues. We had come quite far beyond an "over-the-wall" stage.

The new boss had come to us from a component development organization, where the products were all quite similar and had few parts, but were complex to design and difficult to manufacture with high yield. Suddenly he was responsible for complex consumer telephone products. He was trying to understand how to position the department so that we could make the best contribution, and he asked me how I believed we ought to be organized. Having already spent six years working in the product development environment, I suggested that when one of our physical design engineers was assigned to a project, that person ought to move their desk so that they sat among the electrical engineers working on the project. In that way, the physical designer could be as involved as possible in the day-to-day evolution of the design of the new product, and so could make the maximum contribution.

The new boss, who was used to a tightly knit organization focused on optimizing the design of a family of very similar products, reacted immediately to this suggestion. I remember his gesture clearly: he made a tight fist, and said, "No, we need to stay together."

With his background in a strong functional organization, this middle manager was not ready for multi-function teams. Happily, the complexity of the products and the dynamics of the environment carried him to a point where he eventually accepted less control and more teaming.

Teams as an Important Part of the New Paradigm

Several drivers have recently emerged that lead to a new focus on employees working in teams rather than as individuals. These include a process focus, concurrent engineering, and empowerment of employees.

One driver is the shift in focus to process management that we have been talking about in this book. As you begin to look at activities as *processes* instead of as strings of separate steps, you almost cannot help but see that if the individuals working on the various processes would get together and coordinate their actions, they could probably perform the entire job faster, and perhaps with fewer participants. They would, in fact, be working as a *team*.

The use of concurrent engineering, described in Chapter 7, to create and introduce new products has brought with it a new focus on working across functions in an organization. Employees are asked to come together to work product realization as a team. Indeed, if you have told your design engineers to "get with" their manufacturing engineering colleagues at the start of a new project, you may even have been using *multi-function design teams* without realizing it.

The increased value placed on *empowering* employees as a means of improving productivity while simultaneously improving employee satisfaction also leads to increased reliance on teams. The premise is that a group of empowered employees working as a team, depending more upon each other than upon supervision, will have ideas, energy, etc., that are greater than the sum of the individual contributions from the people who make up the team.

The combination of these drivers—process management, concurrent engineering, and empowered employees—has created an environment that is team based, and making these teams effective in a traditional R&D environment is one of the challenges that everyone who undertakes this journey must work with.

It is very likely that your organization already makes use of several types of teams. Perhaps product teams manage each of your products from "cradle to grave." Product development usually has one representative on such a team, along with representatives from other parts of the business (e.g., marketing, manufacturing, sales, and service). Within the development organization you may utilize a project team to carry out each specific development project, with a project leader coordinating representatives from all the development functions (hardware and software design, system assurance, etc.).

In this chapter we talk about ways to make increased use of teams to support the drivers identified in the preceding paragraphs, thereby improving your ability to produce *Quality products* in a *shorter interval* at *lower cost*.

Much of the material that follows is most directly applicable in large organizations, with many functions that may have heretofore been operating essentially independently, and with many people. Actually, all but the smallest companies can probably effectively use the new focus on teams to improve operations. Even when some functions are represented by only a single person, the traditional way may have been for each of these people to strive to make their function look best, rather than focusing on the best overall outcome. Emphasizing teamwork can perhaps help there. In a small company tightly run by the owner, what might happen to productivity if the owner could be persuaded to take the risk to give up old controlling habits and let a self-directed team do the work? These are among the ideas discussed in the pages that follow.

Benefits of Increased Use of Teams

By making increased use of teams we are able to get the most out of the processes we use and the people we employ. The most common measures of the success of this are these three outcomes, to which we have referred several times:

- Improved product quality, achieved in a
- Reduced interval for developing the product and introducing it into manufacture, at a
- Reduced development cost

Our first attempt at a multi-function team including both R&D and Manufacturing clearly achieved the second and third of these.

Example. Drafting is a significant part of the cost of a development project. Despite the advantages of computer-aided drafting of electronic circuit board layouts, re-doing a layout to accommodate design changes takes time, costs money, and risks new errors. Our first multi-function team that brought together the product development and manufacturing engineers was challenged to improve upon our average of three design cycles to get a circuit board layout that was ready to manufacture in quantity. In fact, they were challenged to do the job in a single cycle (we'll look at more of the implications of this later).

They achieved this goal—in part because the manufacturing engineers were full partners in the development effort, and worked with the rest of the team to ensure that the first board layout was ready to manufacture. Being unfamiliar with the process that the development people followed to design a circuit board, the manufacturing engineers also challenged some of the "usual" ways of doing things, which also helped achieve the goal on a single design cycle.

Some other benefits from the use of teams include:

- Better utilization of employee abilities and intelligence. Employees working together in teams—especially if they are empowered to find their own solutions—are usually energized to do a better job. Experience shows that in this environment, people work together and share the work load, so that they add dimensions and variety to the job.

- Members of empowered teams tend to develop a sense that they *own* the work and the outcome. This energizes them even more, with often dramatic results.
- It has been suggested (Wellins 1992, 24) that innovations to reduce defects come not only from individual "super leaders" but also from the body of workers, working together.
- "Working in teams makes it easier for people to communicate and provide feedback, so it is simpler to make improvements" (Drake 1992, 40).

You do not want to go to your own leaders with only these "arm-waving" claims about the benefits to the company from making more use of teams, so let's look at some real data.

Cadillac's use of multi-function teams to improve the quality and rapidity of their product developments has been widely publicized since they won the Malcolm Baldrige Award. The empowered teams established as part of Cadillac's *Simultaneous Engineering* program are "compressing the time needed to bring new models to market by eliminating waste caused by lack of communication and [lack of] teamwork" (McCusker 1992). The 1988 Eldorado is cited as a good example of how the process works. This product was brought to market two years earlier than would have been possible using the traditional methods.

David Webb reported in *Electronic Business* that, using empowered teams, Hewlett Packard's Disk Mechanisms Division in Boise, Idaho, created a new OEM disk-drive unit in "just eight months, from concept to production. . . . The disk-drive industry average is about 12 to 18 months" (Webb 1992, 55).

Brooktree Corporation, a San Diego-based supplier of mixed-signal ICs, reported to Webb that "a traditional serial approach would take us two years. Our 482 product, by comparison, took us just nine months from product definition to production" using development teams that are "empowered to run their projects" (Webb 1992, 57).

As mentioned above, an empowered multi-function (product development plus manufacturing) team in the authors' organization at AT&T in 1992 brought the hardware for a new business telephone control module from concept into production using a single prototype cycle, compared to an average of three prototype cycles for the organization's most recent similar projects.

Types of Teams We Will Discuss in More Detail

Four types of teams fit into the context of this book. These will be discussed in more detail in this chapter and the rest of the book, and are:

- Multi-function teams to develop products
- Single-function teams to manage functional processes
- Multi-function teams to manage process interfaces
- Short-term teams to manage specific process improvements. These may be single- or multi-function, depending upon the improvement to be worked.

We look at each of these teams in turn, and see what is special about them.

Product Development Team

A multi-function team formed to develop a product typically brings together the design engineers from the product development organization and manufacturing personnel who will put the new product into initial production. All key development jobs should be represented *from the start*, including systems engineering, hardware and software design, system assurance, and field support. It is important to move beyond the old view that functions need only to be represented on the team when they are executing their part of the development process. System assurance can start developing their testing approach while requirements are being detailed, and field support can often add valuable insight to the architecture. Manufacturing people who might serve on such a team include a factory test engineer, a manufacturing process engineer, a material manager, and a purchasing agent. The responsible product manager from the marketing organization may also serve on the team, to represent the ultimate customer and the requirements to which the product must be developed.

Process Management Team

A good way to document, manage, and improve a particular process, such as that used to develop software, is to form a *Process Management Team* (*PMT*) from the community of users of that process. As users of the process they know its good points and its bad points, and can recognize process improvements that will make their own work easier. They understand how to measure the performance of the process, so they can clearly identify the gains achieved by improving the process. Each major part of the process (each "sub-process") would be represented by a PMT member, who takes responsibility for keeping the documentation for that sub-process current, for investigating suggestions for improving it, and for modifying it to incorporate agreed-upon improvements. The PMT as a whole ensures that the sub-processes work smoothly together, and that the interfaces between their process and other processes are made to work smoothly.

Product Realization PMT

Just optimizing individual functional processes will not assure you of the best set of processes, capable of being used to develop new products in the shortest possible interval. It is also necessary to pay attention to how the functional processes interact. Are all the inputs needed by the development processes provided by the outputs of the product planning process? Does the process used by the hardware developers provide the test bed needed by the software developers at the time they need it? Another type of Process Management Team, composed of representatives of each of the functional processes along the path of your product realization process, can address these issues and can sponsor improvement projects for optimizing the ways in which the functional processes interact. Systems engineers will probably play a major role on this team.

Quality Improvement Team

Each of the three types of teams discussed so far is typically a standing team, expected to stay together for an extended time. The product development team will normally stay together until the product has entered standard production. A PMT may last even longer, although individual members may come and go due to staffing changes and career development moves. Another type of team that is useful in the context of this book typically has a much shorter life: a team that is created to address a particular process improvement need or opportunity. Usually made up of users of one functional process, where the process improvement opportunity exists, a team like this may be called a *Quality Improvement Team*, or *QIT*. Some people refer to them in terms of the improvement project they are brought together to undertake, calling them a *Quality Improvement Project*, or *QIP*.

Of these four types of teams, a product development team probably will work the closest together for the longest time, with the greatest potential for working under pressure. Because of this, most of the following discussion relates to this type of team. Where the differences are most important, specific reference will be made to the other types of teams.

Each of these types of teams can be led in either of two ways: either as an "old-fashioned," outside-led team, typically led by a manager above the team in the corporate hierarchy, or as a *self-directed* or *empowered* team, where the leadership is part of the team. These days an empowered team is still something very special; you are likely to require some adjustments in the organization and the work environment in order to obtain the most benefit from the work of such a team. We first discuss things that are common to both "led" and "empowered" teams, and then take a separate look at empowerment and its potential benefits.

THE TEAM ENVIRONMENT

Team Membership

People in a Team Work in Special Ways

A group of individuals who are told to work together do not necessarily and instantaneously become a "team". "Teams differ fundamentally from working groups because they require both individual and mutual accountability. . . . A team's performance includes both individual results and . . . 'collective work-products'" (Katzenbach and Smith 1993, 112). Indeed, "Groups that become real teams seldom think of themselves as a team because they are so focused on performance results" (Katzenbach and Smith 1993, 117).

Exactly how team members work together is determined by the sort of team they form and the job to be done. Just as the members of a basketball team interact differently than do the members of a baseball team, so do the members of a manufacturing line team, who, when work is flowing along the line, may function almost as though they were a "well-oiled machine," interact differently than the members of a quality improvement team, who may bring their separate skills independently to bear on the problem at hand. (This analogy to sports teams is suggested in Torres and Spiegel 1990, 15.) Understanding the nature of the job the team is formed to do is thus important to understanding who is needed on the team and the specific roles that team members will play.

Just telling a group of people that "you are now a team; go to it" does not ensure that they really will "go (right) to it". Two steps that help a lot are for the new team to go through some sort of *team-building* activity together, and for the team members to share formally some goals or objectives.

Team-building activities can vary all the way from taking a half-day class on team issues to an intense shared experience like "Outward Bound" or "Project Miracles."[1] The choice depends upon the type of team and the number of people involved, and also upon the time and funds available for this purpose.

1. "Project Miracles" is a program owned and offered by Sportsmind, Inc., 83 South King Street, Seattle, Washington 98104.

People in our organization starting up product development teams have found it very useful to hold a two- or three-day off-site *Value Engineering* meeting. This combines team building with getting a good handle on the most important features of the proposed new product and their associated costs.

No matter what approach is taken, it seems to be generally helpful to engage one or more outside facilitators who have team-coaching skills (and perhaps psychology backgrounds), to support a team-building activity. *Outside* does not necessarily mean "outside the company," although this may be very useful; resources of this type are often available in company training, organizational effectiveness, or human relations organizations.

Ongoing activities that help strengthen the relationships in a team include regular meetings, working at least part of the time together in a laboratory or conference room, going to lunch together, and ordering in pizza on a Friday afternoon or during a late-night work session.

The members of every team share some sort of objectives—that's why they are a team. Formalizing a set of goals or objectives can also strengthen the team relationships. Recall the example given in Chapter 7 about building the relationships between our development and manufacturing engineering organizations. The managers of both groups wanted to make clear to the associates, and the people with whom we all interacted, our commitment to working together as a single team (here the term is used in a rather loose sense; very directed project and process teams were later formed within this larger community-like team). We drew up a "Relationship Value System," which we all signed and then distributed to our people and to our coaches. Figure 7.10 presents what we wrote; the parts to focus on at this point are:

> *The right design and manufacturing people put together in a productive team environment will make the right decisions, which benefit the business as a whole. Our ability to* ***perform as a team*** *in managing change and introducing new designs is critical to our success . . . Teams will perform in accordance with how they are evaluated. The performance of our* ***teams*** *in these areas and the quality of their teamwork shall be the predominant measures of our success in managing change and introducing new designs, and will be the guiding values in all evaluations. Outstanding examples of teamwork will be recognized.*

The meaning was clear: the coaches from the heretofore separate development and manufacturing organizations had committed to support the people working together as one big team, and to share these goals. This statement formed the basis for a new way of working together, and for a lot of trials—mostly successful—with teams of various types in the months that followed.

It is interesting to note that this was signed and distributed a year and a half before the company instituted a formal change in performance evaluation that accounted for teamwork and "living the values." A couple of lessons can be extracted from this:

- In this new environment, enterprizing leadership does not need to wait for the bureaucracy to act, to move their own organization toward new ways of working.
- This can be true for leadership *at any level*, for the part of the organization they lead.

To make sure that we did not lose sight of these shared values, we made up an overhead to be used as a "focusing" visual, to be on-screen at the start of every meeting we held together. As shown in Figure 8.1, it embodied the idea that with our joint team as the foundation, our shared values supported our customers.

These issues are *very* critical for self-directed teams, and they will be considered further later in this chapter.

Team Roles

Each participant on a team will have a role to play. These three general roles will very likely be present in all teams:

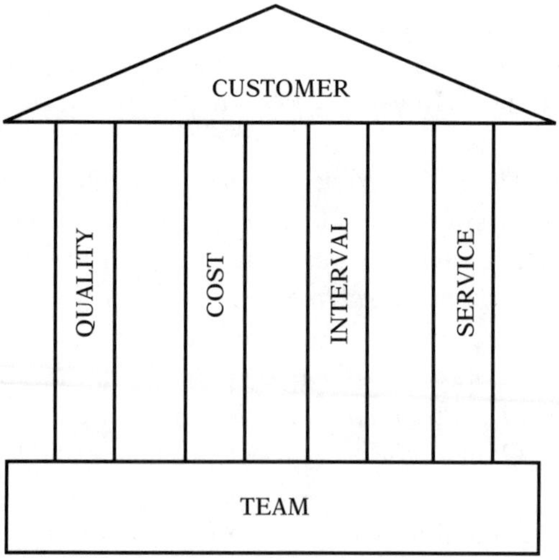

FIGURE 8.1 *Visual We Used to Continually Remind our Joint Team of our Shared Values*

- Team members
- A team leader
- A team coach

For many people, the first time they serve on a team these roles will be new and unfamiliar. It will help to have a clear set of "Roles and Responsibilities" defined, along with clear expectations of how they are to fill their roles. The more a team is empowered to act on its own, the more important clear understanding of roles will be. We briefly consider each of these three roles.

Team members are *all* the people participating in the team's activities. These are the *doers*, who make things happen so that the team achieves its objectives. Their special responsibilities are to apply *all* of their skills to the problem at hand, and to value the success of the team above their individual success. They must commit to the team sufficiently to believe that their own success will come from the success of the team as a whole. This will often lead to each team member's using not only the special skills for which they were chosen to be on the team, but *all* of their skills, as the team members work together to achieve the team's goals.

Usually there will be a *team leader*. This is the person who typically brings the team together, presides at meetings, mediates disagreements, and interfaces with other parts of the organization. There are some special cases of small teams or teams with special missions where there will not be an identified leader; the details of these are beyond the scope of this book. Often the leader will be a member of the team (this is especially true for self-directed teams), but many teams, especially in the early days of using teams in an organization, will be led by "managers" identified to fill this role. This is discussed in greater detail in the "Team Leadership" section.

An evolving role is that of the *team coach*. This is a person who, for example, advises or coaches the team when they have problems working together as a team, or understanding how to interface with their individual managers, or understanding just how much latitude they have to "buck the system" to get a job done. For a team with a leader who is a team member, this is the usual role for the management employee who has a job to be done and who has formed the team to carry out that job. For a team with an external ("management") leader, the coach may be the team leader's supervisor, although this is not necessary.

A fourth role that may prove useful, especially when process focus and teams are new to an organization, is that of the *process coach*. This is a member of your quality or process organization who has the responsibility to help the team to get organized, develop the needed team skills, understand how most effectively to follow the processes available to them, and get started on their job.

Teams and the Hierarchy

Who fills what role should not be determined solely by organizational level. Skill type, skill level, and access to needed information are more important reasons to select people to be invited to join a team.

Obviously this must be tempered by the existing organizational environment. We can envision a day when this can be absolutely true for the members of a team. Today, however, it may be impossible for the nonmanagement engineers on a team to accept that technical ideas given them by a fourth-level manager are "only suggestions" based upon the manager's umpteen years in the business. The other team members may find it very difficult to see this person without seeing the "mantle" of the high-level manager. (The fourth-level managers may also have trouble understanding why their suggestions are not immediately adopted. This is when some support and guidance from top management is needed . . .) On the other hand, it may be very much worthwhile doing whatever can be done to move closer to the ideal, in order to benefit from the experienced person's knowledge of both what can and what probably cannot be done. This is also a good way to start changing the environment.

The team's leader ought not to be a certain-level manager just because this is "who it always has been." The emphasis must be on *who can get the job done*, rather than on what level manager traditionally did this. This topic is discussed in more detail in the section on "Team Leadership."

Other Interactions

In the team's local environment, team members will most likely be interacting with people filling these additional roles:

- The *project leader* of the larger project on which they are working, especially if this individual controls the project funds
- The *process owner* and *process team leader* of the functional process they are using to do their work
- Each team member's *functional manager*

The team members will need to work out how they will work with each of these people. Some of the guidance in this chapter can be applied; part of what

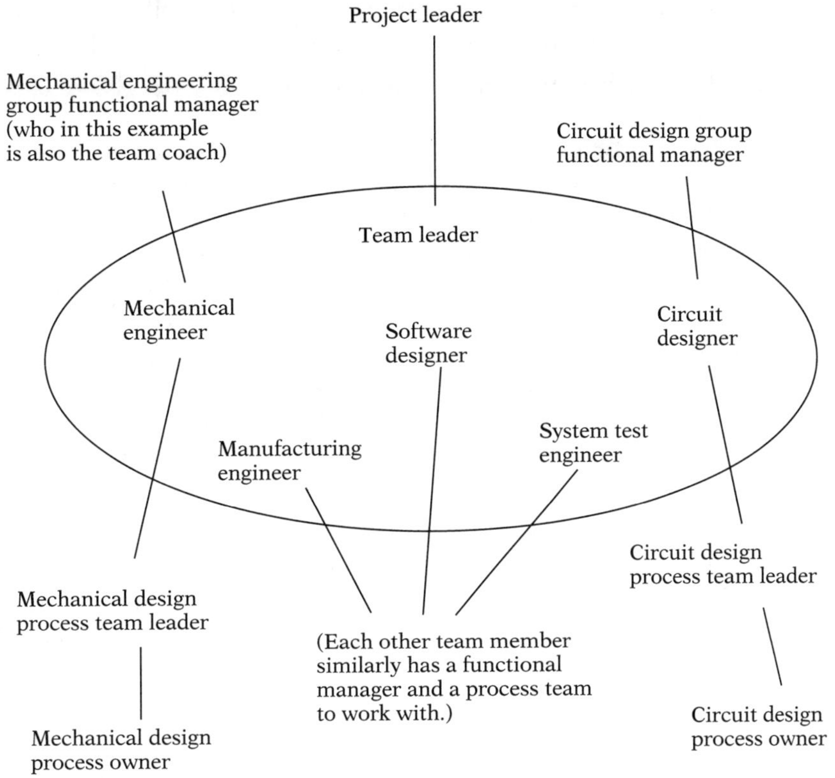

FIGURE 8.2 *The Interactions of Team Members with Others. (The Team Members identified in the figure are examples, as are the Leader and Coach relationships shown.)*

will need to be worked out are the person-to-person relationships. The team coach or process coach ought to be able to help out here.

We've described quite a few interactions. Figure 8.2 shows schematically how the team and team members interact with different people.

The team leader will probably have the most interaction with the project leader, although everyone on the team may at one time or another need to interact with the project leader—at least to have some expenditure approved.

Each team member needs to bring the best possible functional process to bear on the project, so each individual needs to stay in close contact with the people who are overseeing the management and improvement of that functional process: the process owner and/or the process team leader. This should be a very "helping" interaction. There are also times when a team member will become involved with improving the process, either by serving on the PMT or by participating in an improvement project. This will require time balancing, and will need the support of the team leader and the rest of the team.

A team member's functional manager needs to provide a supportive environment, and should focus on the person's ability to contribute, through coaching, education and training, and career development. In a team environment, the functional managers must really be *coaches*, not managers in the traditional sense. They do not *manage* the work of the people in their organizations, they *coach* them to ever greater achievements. They do not tell their people *what to do* or *how to do it*; they help them to find resources and training, and guide them to use the methods and tools of their functional specialty.

The functional manager may also be involved with managing the functional process the team member is using, perhaps as process owner.

Roles are discussed at greater length in the section on "Self-Directed Teams." In that environment there is even greater need to ensure that the roles are well defined, and well understood by the people with whom the team interacts.

Full-Time vs. Part-Time Participation

This is a question that must be dealt with right up front. We have already considered it in the context of concurrent engineering teams in Chapter 7, in the section on "Team Formation and Start-up."

You will probably hear from some quarter the claim that people with two "bosses" (a "regular" manager and a team leader who is a manager), when faced with a priority call for the use of their time, will choose in favor of the manager who evaluates them and gives them their raises. Like many folk tales, this has a lot of truth behind it. People usually do behave in ways that will maximize their rewards.

Consider the dilemma faced by a person working on two teams, when both teams encounter crisis situations at the same time. How does the individual choose where to focus their attention and energy?

Both of these potential situations lead to the conclusion that the key people on a team need to be full-time participants. These are the people whose major work is on the team's project, and who usually must concentrate on a single project because of the scope and complexity of their contribution. Each person's "regular" manager knows that the individual is committed to the team, and the individual has no conflicting demands on their time and attention.

There will be other people needed for a team whose contributions may be relatively small, especially at certain periods in the team's work, and whose usual modes of operation are to share their time among several projects (purchasing agents often fit into this category); these people will have no problem working part-time with a team.

The choice of full-time versus part-time depends upon several factors, including:

- What type of team it is
- The nature of the team's job
- The individual's role on the team
- Where along the path to completion of the job the team is

Development teams, which have a very focused job to do in a well-defined time frame, usually need at least a core of full-time members. A process management team may at most have a full-time leader, if the process is one of the key processes in the organization. A Quality improvement team will usually be only a part-time job throughout its limited existence.

The "core" people on a development team who need to be full-time will be those whose skills are most needed to complete the job. A team developing an electronic circuit board will need full-time circuit designers and full-time mechanical designers, but will probably not need a full-time shipping carton design engineer. In contrast, a team developing a hydraulic test facility will need full-time mechanical designers, but may only need the part-time involvement of an electronic controls specialist.

Early in the work of a development team there may be several design engineers but only one full-time manufacturing engineer, but as the product nears full-scale manufacture, some of the design engineers may move on to new projects, and manufacturing process specialists may join the team.

Team Leadership

Some points have already been made about team leaders:

- Usually a team will have a *team leader.* There are some special cases of small teams or teams with special missions where there will not be an identified leader; the details of these are beyond the scope of this book.
- The team's leader ought not to be a certain-level manager just because this is "who it always has been." It is true that especially in the early days of

using teams in an organization, many teams will be led by managers identified to fill the role.

- Often the leader will be a member of the team; this is especially true for self-directed teams.

While there is great value in having a team led by a person who can gain easy access to other managers who may need to be engaged in order for the team to work most effectively, there is also great value in having the team leader be a capable person at the lowest level that will enable the team to function effectively. Such a person is most likely to be close to the issues that need to be decided and to the human issues that affect the team members, and also less likely to be viewed by the team as "a big shot who is here to tell us what to do." So even if the team leader is a manager who is not a working member of the team, that person ought still to be the lowest-level manager who can effectively lead the particular job and team. This is consistent with trends to move decision making to the lowest possible level, makes good use of employee skills, and may help the team to work more comfortably together.

On the other hand, if the team leader lacks the experience, networking, and span of influence to get the job done effectively, then the team, the job, and the leader all may suffer. We have seen a case where a low-level manager was made the project leader for a major development project, the scope of which turned out to be too big for him. It became necessary for a more experienced, higher-level manager to assume the role of team coach in order to overcome the larger obstacles effectively.

Identifying the Team Leader

The team leader, identified when the team starts their work together, may be designated by some manager or may be chosen by the team members. This can be true even when the leader is a manager who is not a team member; in this case the team might ask the manager who forms the team to assign as team leader someone with whom they know they can work especially well.

Particularly when the team leader is a team member, the leadership of the team may remain with this same person throughout the entire project, or it may move around as the project progresses. In the latter case, the leader at any time may represent the technical skills most important at the particular stage of the project. The leadership role might, during the planning stage, be held by a marketing or system engineering person, then move to a development person during the peak design and development effort, and then later move to a manufacturing person. If such a rotating team leadership is permitted to occur or is encouraged, there probably will not be need for a formal procedure—the appropriate person will very likely emerge or arise at the appropriate time.

It is important that the team leader also have better-than-average *human* skills—leadership skills, interpersonal skills, etc. Look for these skills even when the leader is not someone who has already been recognized for these skills by having been promoted to the management ranks.

The Team Leader's Coach

The team leader's coach (functional manager) will very likely have some special problems with which to deal. To start with, it is very likely, especially in a situation where the leadership of the team changes with time, that the leader's coach is not the project leader. In such a case, the team leader's coach must be careful to really be a *coach*, and work very hard not to try to control or direct the work the team leader is doing. This can be very hard at first, but once the relationship is working well, it feels very good to both parties!

One author recalls his first experience at being the coach of the leader of an empowered product development team. A committed and experienced designer, he found himself in conversations with the person he coached, wanting desperately to leap to the white board and start to sketch his ideas of how to execute the design on which the team was working. It was terribly hard, both figuratively and actually, to keep his hands in his pockets when talking with the team leader. But it was necessary, lest the team leader lose the sense that he and the rest of the team really were empowered to do the design. The coach needs to walk a fine line, and develop the relationship so that wisdom and experience can be passed on without creating the sense of the coach telling the team leader (or other team members) what to do.

Direction from Outside the Team

When the manager for whom a team is carrying out a project is not a member of the team, and also is not the team leader's coach, this person, too, has a unique job. They need to be *sponsor* of the project (concerned about its success), *empowerer* of the team (even when the team is not strictly a self-directed team, it is still the project sponsor who gives the team the go-ahead to do the work, and who must keep the team members free of obstacles so that they can continue to do the work), and *coach* of the team as a whole (keeping them motivated and pointed in the right direction, helping to build skills, and removing obstacles).

In the first blush of working as teams, some team members and (especially) leaders may want to reject outside direction. This is especially likely in the case of a self-directed team. We have seen teams develop the sense that once they have been launched, they are really *on their own*—essentially independent. Hold on! There is always someone who has asked that the job be done, and caused the team to be mustered to do the job. Without this person and their objectives, there would be no project, and no need for the team. Watch out for those who say the team should not take direction from outside. Katzenbach and Smith stated the real situation very clearly: "Building ownership and commitment to team purpose . . . is not incompatible with taking initial direction from outside the team. . . . Most successful teams shape their purposes in response to a demand or opportunity put in their path, usually by higher management" (Katzenbach and Smith 1993, 112).

Another factor to note in assessing the level of coaching and direction that a team needs is the team's *maturity*. A newly formed team, or a team with several members who are new to working in teams, will probably need more coaching than a team that has been successfully working together for some time.

Tools for Teams

Teams can be helped to do their work by a variety of tools. Actually, everyone probably needs the same tools, but teams find them especially useful to working in their new mode. Communications among team members are enhanced, and there are fewer nitty little in-process questions to be dealt with.

A *Readiness Assessment*, most likely administered and interpreted by a member of the quality or process support group, will help a team starting out on a new project to understand if they lack any needed skills, tools, or support. Making sure to correct any such lacks before the team gets deeply into the project will help prevent the team's coming to an unnecessary and schedule-devastating halt later on. An example is given in Chapter 12, "Getting Started" (see Figure 12.3).

Processes that are *documented* and proven to work help the members of the team get pointed in the right direction, and help them with a lot of along-the-way questions of how to proceed. An expert working alone will often make up for the lack of a documented process by essentially defining the process as the work progresses. If a team tried to do this, they would probably lose time debating about whose vision of the process is best. Having a documented process to guide them eliminates this problem. Ideally, the team will also accept responsibility for initiating improvements to the process as they encounter opportunities to make it even better.

Templates for things such as common schedule elements (e.g., "circuit board layout" in the electronics design business) help in much the same manner as documented processes: the team does not have to spend time figuring out or debating among themselves what the basic schedule activities, milestones, or intervals should be.

Schedules themselves are especially important tools for teams. They not only enable planning the work and tracking the project (which they do for anyone working on a project), they also help the team members keep synchronized. Without a good schedule agreed to by the whole team there is opportunity for lots of day-to-day jockeying over what needs to be done next.

Other *Project Management Tools* that are shared by everyone in the team can help in a similar way. If each team member can enter needs for change into a common change control system, and then the team meets together at some appointed time to discuss and resolve the collected change requests, the team can share the benefits of keeping moving ahead and at the same time share in the decisions of what to change and what not to change. Clearly documented agreements with suppliers, customers, and the project leader, negotiated up front, can save time and confusion as the project progresses.

If every member of the team is using the same *Software Applications* (e.g., word processing, spreadsheet, presentation, and project management software), communication is significantly enhanced. Everything from technical data through documents to schedules can be readily shared, whether it is by exchanging disks or by using a network.

Rewards and Teams

"Ultimately, however, the satisfaction shared by a team in its own performance becomes the most cherished reward" (Katzenbach and Smith 1993, 119).

True as this may be, you may also want to develop new ways to recognize and reward teams. A very special need is to reward for team success, not just for individual success. This can be very difficult within the context of performance review, especially at the start. This is discussed further below. Rewarding team performance is less complicated when it comes to other forms of reward and recognition, some of which are described following the discussion of "Performance Review."

This is a sufficiently important subject that the literature on it will probably continue to develop significantly.

Performance Review

One area to which you will need to pay close attention is how your existing traditional reward structure supports teams. In many companies, especially in R&D and development, people have for years been hired and rewarded based upon their abilities as individual contributors. It is a shock to both the individuals and the system to suddenly change the focus to teams. You will need to understand how your company's existing performance evaluation processes work in order to decide how they need to be changed to recognize contributions to process and quality clearly, and to recognize contributions by individuals as members of teams. It may not be possible or desirable to move immediately to an environment where people are not directly compared, as suggested by Dr. Deming (Scherkenbach 1986, 47), but it will also no longer suffice that the only input on an individual's performance comes from the person's immediate functional manager/coach. Where the person appears to have "two bosses," their functional manager and the team leader, it is absolutely critical that the team leader's evaluation plays a major role. The functional manager *must* solicit this input. You may also want to modify the performance review process to include input from all or some of these people:

- Team colleagues
- Team coaches
- Managers/coaches of other team members
- Project leaders
- Members of the appropriate process management team

This has been termed "360° feedback" since the performance input is gathered from people on all sides of the individual.

Other Forms of Reward and Recognition

Among these special rewards may be special recognitions, events, and distinguishing paraphernalia customized to the team. Special recognitions may be as simple as posting a photo of the team on a bulletin board in the organization's common area, or ensuring that the team's existence and achievements are recognized in the organization's newsletter. Events could include team lunches when significant milestones are met, or a visit to the team by a company leader they don't usually get to meet. Customized paraphernalia could include paper weights, polo shirts, or magnets decorated with a logo or a slogan dreamed up by the team to represent their special work mode or some characteristic of their project. Figure 8.3 is an example of such a magnet; more about this logo, including its meaning and its origin, can be found in Chapter 9. This Team Award is intended to be given out by team leaders or coaches when the team or some team member has done something of special note; it

FIGURE 8.3 *Team Award Magnet*

is a "low-energy" award that does not require extensive deliberation or approval.

This has been only a partial introduction to the concept of new kinds of rewards and recognition. The sky is, more or less, the limit for creating your own—within the context of the organization's culture and the available funds. Our bias is toward doing more to recognize and thank people, rather than less. You may find useful ideas in the extended discussion of rewards and recognition in Chapter 14.

Special Rewards for the Process Team
There may also be a need to find some special ways to reward your process core team, especially if the company tradition has given (and still gives) the most credit to those who contribute directly to product development. Any of the sorts of special recognitions suggested for development teams also apply to this special team. Recognition by the project teams they are serving may be the most meaningful expression of appreciation to these team members.

SELF-DIRECTED TEAMS

The Value of Self-Directed Teams

It has been said that "Self-Managed Work Teams are the most advanced form of employee involvement" (Yeatts, Stanley-Stevens, and Ruggiero 1992, 26). The benefits to the company can be dramatic:

> *Organizations that use self-directed teams have maintained or reduced labor costs while improving productivity by fifty percent or more* (Wellins 1992, 26).

There is clearly a growing collection of experiences and data that lead us to believe that organizing people in *self-directed* or *empowered* teams is an incredibly powerful way to improve your organization's performance and results.

When you think about it, having teams of developers make their own decisions makes a lot of sense. Considering how much engineers are typically paid, can the company afford not to have them make decisions? Are they paying their way if for every decision they must go to a "boss"? There are clearly some decisions, especially those involving significant investment or great risk, where experience and wisdom must be factored in. But there are many development decisions, such as making minor component changes to increase the robustness of a product in production, where requiring a management approval signature on the drawing is really not necessary. No value is added—which in modern economic times should lead us to ask why the approval step cannot be eliminated. This is the first step toward creating self-directed development teams.

Roles in Self-Directed Teams

Functions

Of course, the primary role each member of a team fulfills is to do the functional job for which they were brought to the team. With time, as the team members get used to working together and cross-train one another, individuals may do some of the functional jobs that originally "belonged" to others—but each person's application of their own skills and expertise and experience to achieving the team's goals is primary.

For this reason, it is key that the people recruited to a team have high skill levels, and that their skills are continually enhanced by means of ongoing training. Without highly skilled team members, the other benefits of self-directed teams can never be fully realized.

Roles

The principal team roles were identified earlier:

- Team members
- A team leader
- A team coach
- Process coach

Especially when self-directed teams are new to an organization or to the members of a particular team, the role of process coach is important.

The first time people serve on a self-directed team these roles will be new and unfamiliar—even to people who have been members of traditionally managed teams. It is critical to have the roles and responsibilities clearly defined, along with clear expectations of how people are to fill their roles. These same definitions will also help people outside of the team to understand the behavior they are observing in the team.

As was stated earlier, the team members will also be interacting with people filling these additional roles:

- Project leader (the team's "customer")
- Process owner and process team leader
- Each participant's functional manager

Team members' interactions with the project leader and process owner will be essentially the same as described above. Their interactions with their individual functional managers may be significantly different in a self-directed team environment. The functional manager really needs to serve as a coach to the individual: ensuring that skills are kept sharp, making their own technical experience available in an advisory mode, being a willing listener, providing support and guidance, etc.

To get people to accept these important role differences, you will want to publicize them well. Remember—the people are used to behaving in the "old" ways, to being "managed" rather than "coached" and "empowered". The authors made use of a "Roles" memorandum, which we made available to everyone engaged in empowered teams, and also to all members of management, so that they would understand how to interact with people behaving according to the new team roles. Wide distribution of this information is appropriate, since even those people not immediately involved in the new type of teams will eventually get their chance.

We had some early experiences where participants in team meetings had difficulty when some person present was filling an unfamiliar role. Was this particular manager present as a *coach*, as a *team member*, or as the *team leader*? An artifice developed by one of the authors was the use of colored hats, each embroidered with one of the key roles. With the manager in question wearing a *team member* hat, the other team members could work with that individual as a technical peer, rather than in worker/manager relationships. Considering how (un)carefully a lot of memos get read, the people with an immediate need to understand and work in the new roles can also benefit from training. You may want to hold training sessions for the several roles, and also for each newly formed team, together as a team.

AT&T Industrial Arts—Middletown (Brigitta Hanggi)

FIGURE 8.4 *"Role" Hats*

If you find that some middle managers have difficulty adjusting to the new environment, it may be necessary to get the top manager to cascade the concepts and role of self-directed teams through the organization, so that it is absolutely clear to everyone that this is the new way of working.

Team Member

Each team member of a self-directed team has some special responsibilities beyond the basic responsibilities of a member of any team. Without a "boss" overseeing the team's work and interactions, it is entirely up to the team members to ensure that the team works effectively. The members of a self-directed team need especially to

- Be conscious of the value of teamwork, so that individuals do not pursue their own agendas or the agendas of one particular manager at the expense of the unity of the team
- Be sensitive to the needs and feelings of the other team members, and be courteous in all communications and negotiations
- Support the team leader by surfacing issues and communicating frequently, so that the leader can properly coordinate the week-to-week workings of the team

- Keep their own functional manager informed of the team's work and progress

Team Leader
Leaders of self-directed teams have some special responsibilities, whether the leadership role is taken on "for the duration," or just for that part of the project's life when the individual's area of expertise is the dominant concern of the team. The leader of a self-directed team is very likely to be chosen by the team, rather than be appointed by the project leader or a higher manager.

The leader of a self-directed team must take the responsibility to

- Schedule and moderate team meetings
- Keep the schedule up-to-date, and interface with all people and organizations affected by any schedule changes
- Track, prioritize, and follow up on action items assigned to the team
- Be sure that the team follows the appropriate process(es)
- Represent the team to the rest of the organization
- Be a role model for the team members, in being conscious of the special obligations of team members, described above
- Support every member of the team, to help them meet these special obligations

Some of the special considerations this leader must address include:

- Must want to be the team leader
- Must have time to carry out the team leader's responsibilities
- Must have good organizational skills
- Must be able to coordinate, facilitate, and manage meetings
- Must have good interpersonal skills (communication, negotiation, persuasion)
- Must be capable of motivating the team
- Must be willing to yield to team consensus
- Must always represent *the team*, rather than themselves, other individuals, or management

Ideally, the team leader should be a nonmanagement team member, chosen because they have appropriate skills and/or is willing to serve the team in this role. There may be times when it is useful to have a manager serve as team leader. The team should agree to this: both to the concept and to the individual who is selected. The team members also need to work out with this team leader how the role will be carried out, so that the leader does not fall back into old ways and try to "manage" the team.

Team Coach

The team coach is the management representative responsible for giving the team direction, encouragement, and support. This is normally the person who "empowers" the self-directed team. Usually the team coach will be a manager at the level immediately above most of the team members.

The team coach plays a large part in ensuring that the team forms smoothly. The coach will usually

- Facilitate assignment of team members
- Call the first team meeting
- Help the team to get organized and under way (unless there is a process coach—described below—to help with this)
- Aid the team in choosing a leader
- Participate with the team in team and process training

Once the team is working on the project, the team coach must

- Provide clear goals to the team
- Assist the team with budget and other resource issues
- Provide technical and work experience insight into issues surfaced by the team, without forcing individual bias
- Anticipate and remove roadblocks for the team
- Always support the team concept

Process Coach

When self-directed teams are new to an organization or to the people who are engaged to form a particular team, there is often value in having a person (ideally, a peer of the team members) who is skilled in facilitating teams available to coach them in how to work as a self-directed team. As a peer of the team members, this person can work with them in ways that a manager could not, to get them to accept the "self-directed" role and to develop the necessary skills and relationships. With a knowledgeable peer there is no perceived pressure or coercion, as there probably would be with a manager trying to help them get started.

This process coach would usually be a member of your quality or process organization, who has both teaming skills and coaching skills, and who has the responsibility of helping the team get organized, develop the needed team skills, and get started on the job.

Functional Managers: New Roles for Old Managers

Earlier in this chapter we discussed the new roles that "old" managers must undertake when teams become part of the work environment. Books have been written about this; a short but powerful guide to this new way for managers to work is Larry Hirschhorn's *Managing in the New Team Environment* (Hirschhorn 1991).

Team members' functional managers who are used to "managing" may find it hard to keep their hands in their pockets as they *coach* their organization members. In a development community, these managers are usually good engineers who have been rewarded for some combination of extra-good skills. We need them for their experience and their expertise, but we no longer want them to "manage" what their people are doing—to tell them what to do and how to do it. We want them to *lead*, and to *coach*, in such a way that we get the most complete contribution from each member of the team. Among the things the functional manager can do are:

- Demonstrate support for the principles of working as a self-directed team—not just say it, but walk the talk
- Help the team members to use the methods and tools of their technical discipline to come up with creative design ideas, and to perform the necessary analysis in support of their design ideas
- Make their technical experience available in a non-coercive way, perhaps by means of directive questioning or by providing carefully selected memos and papers as reference material
- Encourage team members when the job seems extra difficult, or when the stresses of working in a new team seem overwhelming

These functional managers need also to be treated very carefully. As they are driven by the changes around them to change the way they work, they may sense a void—a lack of something in their job they found rewarding. They may genuinely miss being the "lead engineers" or "chief negotiators"—roles now taken on by members of the team. We have seen these effects on our colleagues, when our first empowered teams were set up. Remember the example cited earlier of the first experience of one of the authors, at being the coach of the leader of an empowered team, and how he chafed at not being able to contribute directly to the design—truly keeping his hands in his pockets so as not to start sketching design ideas. Functional managers experiencing these roles for the first time need to be shown, usually by *their* coaches, that their new roles are equally, or even more, important than their old "managing" roles. For first-line managers who have hung on as much as possible to their pre-management technical roles, this may require their taking additional training in the *people* aspects of management. They need to be led to understand that they will find ever greater job satisfaction and reward as they improve in performing their redefined jobs. They will probably also need to be given additional new responsibilities to fill the voids, such as functional process ownership.

These new roles have less "management" content—less close-in, day-to-day direction of what the people do and how they do it. Instead the focus is on *leadership*—setting an example of style and dedication, and showing the way—and on *coaching*—hands-on encouragement, skill building, and asking questions that will cause the team to think out all the issues and move toward the solutions.

As team skills improve, their needs continue to change. "A fully self-managing team still needs a coach at times; even more, it needs a mentor and consultant" (Carr 1992, 46).

Another major job of any coach working with a team, be it as functional manager of a team member or as the project leader coaching the team's project work, is to serve as a mediator between the team and other parts of the organization. This is especially crucial during the early days of a team-focused environment, when many of the others are not fully attuned to the ways in which teams work, and the needs of teams and their members. This "running interference" is necessary for keeping the team free to concentrate on their work.

Training for the Team Leader and Team Coach

These are obviously not easy jobs in this brave new world of self-directed teams. Torres and Spiegel (1990, 36) suggest that the successful leader (and coach) in the team environment must be especially adept with these skills:

- Leadership skills
- Communication and facilitation skills
- Training skills
- Consulting skills
- Problem-solving skills
- Feedback skills

Just as the members of a team must be trained to do their new job in the very best way, so, too, must team leaders and coaches be trained in these skills that are critical to making the team environment work.

What to Do with a Non-cooperative Manager

Ideally, every manager will see the benefits of the new ways of working and become converted. But in the real world, there will probably always be some who don't like what they see, who feel comfortable in the old ways, and who do not want to change. While it is hard to see exactly what to do in such a case, it is easy to say what not to do: you *cannot* allow such a person to remain in a position from which it is possible to undermine the efforts of the rest to bring about change, and to make the team approach work. Some change *must* be made, so that such a person does not undermine the team's ability to do the job that they have been given. This becomes a problem for the non-cooperative manager's coach, as well as for you and your team. To the extent that you can, you must work with this noncooperative manager to bring about change. Getting that manager's coach to support and help you can make the change more likely. In some cases, the coach will have to find a new job for the non-cooperative manager so that the work can continue without interference.

Creating the Necessary Supportive Environment

We have already mentioned some requirements of the supportive environment, including

- The need to have middle management be supportive (and the possible use of cascaded deployment of the concepts of self-directed teams, to ensure alignment)
- The need to find other jobs for non-supportive managers

This is part of working the organizational transition to self-directed teams. It may well not be easy. Carr has written of ". . . the immense cultural strain the successful empowered teams . . . create within an organization" (Carr 1992, 43). "Implementing Self-Directed Teams requires ideological and organizational changes within a company" (Drake 1992, 39). The task can be supported

- By value alignment, either with the organization's published set of values, or a set crafted for the part of the organization introducing self-directed teams (like the set shown in Figure 7.10)
- Through a common language (Torres and Spiegel 1990, 57), so that everyone in the organization understands what is being done, and what the people most involved with the teams are talking about

- By training *everyone* in team skills, not only the people who will immediately be involved in self-directed teams. This develops general understanding of what the team members are experiencing
- By giving the managers/coaches more training in interpersonal skills
- By reinforcing the desired behavior, by publicly recognizing those who become role models.

This can get the people moving toward a *team* focus. To go with this, it is *absolutely critical* that you make sure that the rest of the organization supports the teams—especially top management!

PUTTING A TEAM TO WORK

People have found that a team goes through several stages as they organize and get under way. In his 1965 article on "Developmental Sequence in Small Groups," Bruce Tuckman of the Naval Medical Research Institute identified them in this easily remembered way:

- Forming
- Storming
- Norming
- Performing

(Tuckman 1965)

This is especially true for self-directed teams, and less so for "old-fashioned", outside-managed teams. Let's look briefly at what each of these stages is like, and what care may need to be exercised during each stage.

The Team Forms

A team ought not be formed just because it seems like a good idea—setting the goal "'to implement teams' is an inadequate goal. It will almost certainly fail" (Carr 1992, 43). Rather, a team should be formed when there is a job to be done that can be done better by a team than by assigning one or more individuals to do it. "Teams succeed when they have clear, worthwhile, and compelling goals, and team members know that they are expected to commit themselves to these goals" (Carr 1992, 43).

The job should be assigned to a manager who can and is willing to serve as team coach. This person's first task is to gather the appropriate people. The important considerations have already been discussed:

- What functions and skills are needed to do the job?
- Who should be full-time team members and who should be part-time?

As candidates are identified, it may be appropriate to spend some effort "selling" them on joining the team. Again, this is more important for a self-directed team. You will want to sell

- The project
- The goals (e.g., a short interval, or a new level of concurrency)
- The expected rewards (why this is a special opportunity)

You will want to make clear the expected duration of the job and any other unique conditions, and also address performance review for the team members up front.

As you sign people up, take care: "some employees will be apprehensive while others are over-zealous . . ." (Yeatts, Stanley-Stevens, and Ruggiero 1992, 28).

The Team Storms

For a time after the team is launched they will appear to be thrashing about, as they get used to one another, select and develop their relationships with a leader, and develop their relationships with their coach and their functional managers. Bear with them, and help them with the *people* issues.

Very Important: Don't let this go on too long! All of these relationships need to be developed, but there *is* work to be done. At the same time, be wary of "too much too soon too fast."

This is the time when team building begins; it may be an appropriate time for a formal team-building activity, like a class or workshop or retreat. ". . . extensive research in high-performing groups . . . concluded that spirit was at the core of every such group . . . studied" (Bolman and Deal 1992, 43).

Example. One author went through a very strenuous *storming* experience with a team. The term is well chosen! Four very highly skilled, experienced engineers and the immediate manager of two of them (the author) sought to form a self-directed team to coach a number of newly formed process management coordinating teams, each led by a senior manager. A key question was What would be the role of the manager? Could and should he be a team member, or should he be the team's coach? Originally the manager thought that he should be a team member, because the team appeared to be short-handed for the job to be done. But the manager also felt that he "owned" this job, as part of a larger set of activities for which he was responsible. This led to a critical conflict over how the manager would relate to the rest of the team members, should he try to be a full member of the team. How could he *share* the team's empowerment at the same time that he felt that he "owned" the job they were doing? How could he decide what parts of the information he had about his total set of jobs were relevant to the team's activities and should be shared with them? The engineers were concerned that because of this the manager would have information about the job to which they were not privy—which would make them less-than-equal partners with the manager in his role of team member. These concerns were felt most by the two team members who had not already been working with the manager. This, in turn, caused feelings of loyalty conflict in the two who worked for the manager, between their existing relationship with the manager and their desire to be equal partners on the team. Existing, well-working relationships were threatened, and the four engineers were having a really difficult time building the cohesion necessary for them to become an effective team.

The first breakthrough was mutual agreement that the manager should not be a team *member*, but should take the role of *team coach*. Full and open communication was another key agreement. The step that allowed everything to come together was the manager's acceptance that once the team undertook the job, *they*, not he, would "own" the job. He might be the one to report results to the next coordinating level, but the team would assume full responsibility for the job's success or failure. In order that the manager could fill this role and the role of team coach, a process was established for keeping him informed of the team's progress and key activities.

During the deepest part of the storm, there were very serious concerns on all sides. Both the team members and the manager sought coaching (and intercession) from the manager's supervisor. An apparent need to get started on the job added both urgency and stress to the need to resolve the issues. Another source of stress was that the manager and one of the engineers had been building toward this point for a couple of years, and felt a real stake in making the new process management coordinating teams work successfully. The one thing everyone agreed upon was the importance of having this very capable team undertake the job. Largely because of this, they kept working the issues. The

> result was as good as the most intense part of the storming was bad. The team, working with the coach, prepared to work with the coordinating teams, earned their trust, and was able to get them going as well or better than had been planned.

Two important lessons of this experience are that the relationship between a manager and a self-directed team must be carefully worked out, and *storming* will occur, and everyone must resist the transient desire to give up rather than work through the stress.

The Team Norms

Here, as they begin to get under way, is where the team will need support:

- Setting objectives and goals to which everyone on the team subscribes
- Selecting and scheduling training activities to fill important gaps (e.g., specific processes, new tools, and interpersonal skills)
- Arranging for a pre-start assessment (probably through the process or quality support team)

During the *norming* stage, the team may agree to "operating rules" to help members work together most effectively and most comfortably. Here are some example rules from our own experience:

In general interaction,

- Have a bias toward principals
- Live the values
- Feel free to be direct
- Complete homework assignments (i.e., meet commitments)
- Challenge ideas, not the individual with the ideas
- Have fun

At meetings,

- Be on time
- If you send a substitute, prepare that person in advance
- Everyone has an equal voice
- Stay focused
- One conversation at a time
- Have a bias toward consensus

The Team Performs

The rewards begin to come when the team gets it all together, and begins working together to achieve the job for which they were formed into a team. During this time, the team coach, functional managers, and process coach will devote their energies to the responsibilities discussed above under their roles, and also to

- Helping with the team problems
- Evaluating the team's performance
- Continuous improvement

9 Goal Setting for Product Development

Strategic Quality Ideals

✓ Clear Vision and Goals Communicated to All
✓ Customers as the Key Drivers for Change

Organizational Enablers

✓ Leadership and Commitment at the Top
✓ Support and Alignment of Middle Managers
✓ High Morale and Satisfaction of Employees

INTRODUCTION

You probably know that R&D cannot tackle every improvement project or solve every problem.[1] Your product development organization needs to apply a process for intelligently selecting the right areas to work, especially those areas where breakthrough improvements are needed. One process to do this requires that you:

- Create a vision
- Establish long-term plans
- Set annual objectives
- Execute and deploy work plans to achieve these objectives in a Plan-Do-Check-Act cycle; also known as the *Deming* or *Shewhart Cycle* (Deming 1986, 88; Shewhart 1986, 45)
- Repeat the process by re-examining your vision and objectives

This process goes by several names, including *management by policy, policy deployment*, *Hoshin*, and *Hoshin planning* (King 1989, 1–3). One representation of the process is pictured in Figure 9.1.

R&D must step up to such a process if they are to effectively apply their resources to best serve the interests of the company. However, the development organization should not go off and do policy deployment by itself. Policy

1. R&D is used interchangeably with the term *product development* in this chapter as a term to represent the product development organization.

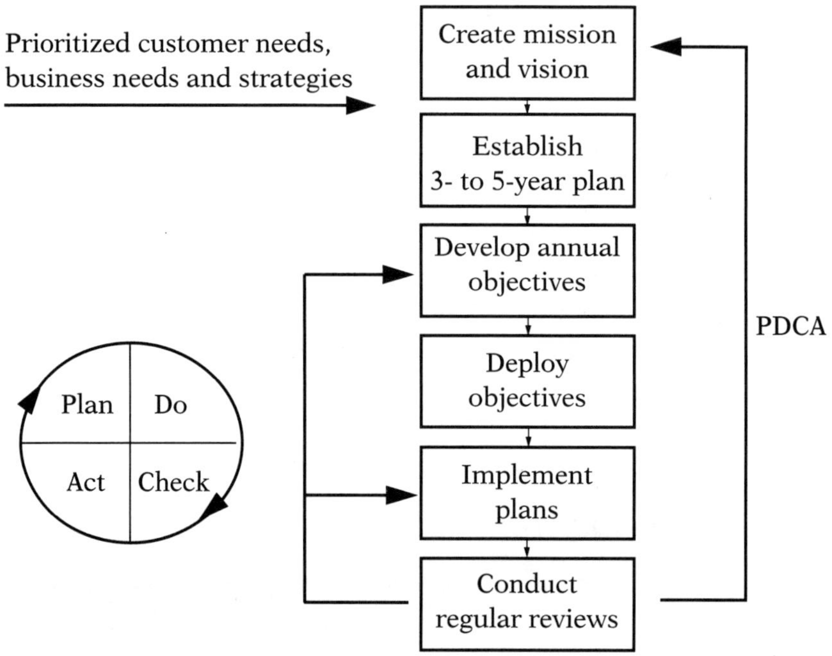

FIGURE 9.1 *Policy Deployment*

deployment should be a business-wide effort to gain synergy between each organization. The right goals for product development will stem from:

- Knowing how product development efforts best align with and support the business objectives
- Continuous focus on meeting customer needs better than the competition
- A clear vision for product development

When the process is executed effectively, each associate should know the key goals for product development, have work plans and clearly aligned objectives, and participate in regular reviews to evaluate progress and act on the findings.

Having said all that, we have observed this activity to be one of the most difficult elements of our quality system. During our careers, we have gone through periods of having clear objectives and other times when we did not.

During the periods that we lacked clear directions and goals, work activities were fragmented and we lacked alignment. Individual managers would self-select the most important areas, and the improvement activities you worked depended on the organization you worked in. The reasons for not having clear goals included lack of buy-in to a goal-setting process by senior management and the sheer complexity of doing this process right. In many cases it becomes such an arduous process that it is tempting to just not do it.

Training and persistent socialization of the benefits and purpose behind policy deployment are effective ways to gain buy-in from senior management. One selling point that you can communicate to your management is that you (your organization) will be able to realize breakthrough improvements only if you focus your resources on a few key objectives. Said another way, you have only limited resources within R&D to execute improvement activities, and these resources need to understand the goals for the organization so that they can apply themselves accordingly.

Helping senior management establish the "right" goals for product development can be challenging. When we attempted to describe the steps in this chapter, we immediately recognized the complexity of the task. In fact, we could not find a way to describe simply the final stages of the goal-setting process where you actually pick the key focus areas; the process does not lend itself to a recipe approach.

What we have elected to do is describe the necessary prerequisites, and discuss in general the considerations that need to be made when selecting and deploying goals and building alignment behind a vision. We believe that the specific goals and directions you choose (as long as you are not too far off base) are not nearly as important as making a choice and moving forward. The following considerations for setting product development goals will be covered in this chapter:

- Customers' needs
- Performance level against the competition
- Business objectives
- The vision for product development

CUSTOMERS' NEEDS

Understanding your current performance levels translates to knowing your customers and your competition. With this knowledge you can begin to plan where you want to go. Recall that one of the Quality Ideals is *Customers Drive Change*. If the competition is gaining ground, then they are doing a better job at meeting the needs and expectations of their customers.

To understand customer needs, you first need to identify your customers. R&D must recognize both external and internal customers—someone other than the end-user of the product. Dr. Philip H. Francis, corporate vice president and chief technical officer of Schneider North America (Square D Company), states, ". . . R&D must understand that it has a customer: the company" (Francis 1992, 19). The goal setters in R&D must understand who "the company" is, and capture the needs of these internal customers.

You can begin to list your customers by writing down the outputs from product development. In other words, what do you produce: drawings, parts list, software code, etc.? In the same straightforward manner you can write down who receives these outputs, i.e., your customers. Table 9.1 shows an abbreviated list of outputs and customers we generated for one part of our development organization.

Associates as Customers of R&D

Product development organizations less frequently identify associates (employees) as customers. Yet consider that associates have both the ability to choose, and they receive an output. Their ability to choose is simply their ability to work elsewhere. Moreover, associates can choose to put their heart and soul into their work, or can do just enough to get along and collect their paycheck. What is the output to associates? R&D managers provide the work environment, tools, processes, educational opportunities, career opportunities, and management coaching to associates. The quality of this output often has an impact on the choices discussed previously. By identifying the associates as customers, R&D leaders can address one of the key Organization Enablers, *High Morale and Satisfaction of Employees.* Although the company pays associates, it is the management of R&D that *supplies* the systems, structure, tools, and overall environment for each R&D associate. This environment, combined with the employee's compensation (paycheck), either generates or destroys morale and job satisfaction.

TABLE 9.1 *Mapping Outputs to Internal Customers for R&D: An Example*

Output	Customer
Printed circuit board artwork and data	PCB vendor
Software	Manufacturing*
Stock lists and assembly drawings	Manufacturing
Piece part drawings for telephone housing	Plastic molding vendor
Drawings and functional documentation	Customer documentation group, services, and training

*In a concurrent engineering environment, manufacturing and services organizations are less customers of R&D and more partners in the development process.

External Customers

Note that all of the customers identified for product development so far are internal customers. They are not end-users. A great risk would be taken if only internal customers were used to develop strategic goals. Locally optimized solutions may emerge that do not serve the end-users of your product. Therefore it is necessary to add to your customer list the end-users of your product.

Recognize the Funders

R&D must also recognize and understand the needs of that organization in the company that chooses *who* will develop new products, and who funds R&D. In some organizations like AT&T, market management (also called product management) makes these decisions. Does that imply that market management or a similar organization in your company is a customer of product development? In a sense, they contract with R&D and manufacturing for a product, and offer that product for sale to end-users or distributors through sales channels. This organization represents the voice of the business to R&D and manufacturing, and can choose *who* develops new products. The internal R&D organization may not always be chosen for a new product development project. Not unlike the way manufacturing can be outsourced (purchased outside of the organization), so can development. The needs of the funding organization are an important part of the goal-setting process, and will be grouped with the other customers of R&D for the purposes of this chapter. We now can write down a complete list of customers. For our organization, a sample list of customers is shown in Table 9.2.

TABLE 9.2 *List of Customers for an R&D Organization that Designs Telephones*

Customer	Type
Market management	Chooses who develops products
Manufacturing	Receives output from R&D
Training	Receives output from R&D
Documentation	Receives output from R&D
Services	Receives output from R&D
Associates	Work in environment created by R&D management
End-users	Receive and use final product and service
Plastics molding and other piece part vendors	Receives output from R&D
Printed circuit board vendor	Receives output from R&D

Understanding the Needs and Priorities of Customers

Once you have identified your customers, you need to gain an understanding of their needs and priorities. This understanding includes an assessment of how well your product or service meets the needs of their processes and operations.

To assess your performance, and to understand what is important to your customers, you need to ask them. An alternative approach is to evaluate your performance internally. The problem with this latter approach is that many of us have preconceived notions about what is important to customers based on our own experience as consumers. We fall into the trap of projecting our values onto our customers. For example, you may be very sensitive to price when you purchase a television set or VCR. You want it to work, be inexpensive, and have a reasonably good warranty in case something goes wrong. Another person buying the same type of equipment may not consider price as their main concern. They may be more interested in the brand name of the company, the number of state-of-the-art features, and the overall performance level of the equipment. Each of you have different values, and because of these values, your purchasing criteria are different. For this reason, rather than presume what is important to customers, *ask them*.

You can gather customer data with surveys or interviews. Which method you use depends on the size of the customer set. In cases where the customer set is small, a survey instrument is neither needed or desired. Direct communication should be used whenever possible. Remember that the goal is to become knowledgeable about your customers' needs. Wherever possible it is best to do this by putting yourself in their place, as if their business or function belonged to you. Your intimate knowledge of customer needs from the perspective of their operations will produce the best directions and goals for you. An interactive dialogue will be much better at getting at the heart of customer needs than will a survey. For large customer sets, a survey is the only practical way to collect input. The survey or interview should assess four areas:

1. The needs of the customer (what they want or expect)
2. The importance of the items in relation to one another (prioritized needs)
3. Information about the customer type to allow categories or segments of customers to be identified (demographic data)
4. Your perceived performance level in each need area

If a survey is necessary to collect customer data, then the survey itself will need to be developed. The easiest way to develop the survey questions is to form focus groups with customers and members of R&D. A focus group is nothing more than a group of people who are representative of the larger population. Work with the focus groups to brainstorm their needs and concerns. Tools like Affinity Diagrams and brainstorming can be used to develop a customer needs list, as discussed in Chapter 6. The results of this focus group session, namely the list of customer needs, can then be translated into the survey. Two examples based on our experience in working with internal customers are shown in Figures 9.2 and 9.3.

Whether the customer set is large or small, or whether surveys or interviews are used, the end result of this process should be a prioritized list of the customer needs, along with the customers' evaluation of your current performance level. For surveys such as those shown in Figures 9.2 and 9.3, these are easily extracted by compiling all of the survey results and listing the top three to five needs. The top three to five needs are those needs that had the highest average importance rating from the survey data. For each of these three to five needs, you will also know your performance in that area from the survey.

Tables 9.3 and 9.4 are examples of what the results might look like for the survey shown in Figure 9.2. All the possible tables are not shown. The demographic data can cause a large number of tables like the ones shown here to be

TABLE 9.3 *Product Management Needs for Telephone Sets: An Example*

Customer Need in Priority Order	Perception of Performance from Customer
1. Fast time-to-market	2
2. Technology platforms available for new product applications	4
3. Skill set matches future product development needs	4.5
4. Able to meet cost targets for new products	3

TABLE 9.4 *Product Management Needs for Telephone Control Units: An Example*

Customer Need in Priority Order	Perception of Performance from Customer
1. Technology platforms available for new product applications	3
2. Able to meet cost targets for new products	2
3. Fast time-to-market	2.5
4. Core research provides new product opportunities	4.5

R&D Customer Survey for Product Management—Example

Directions: Rate each area in terms of importance to you as a customer of R&D, and your evaluation of our performance in meeting your needs in each area.

1. Easy to work with
   Not Important 1 2 3 4 5 Very Important   Poor 1 2 3 4 5 Excellent
2. Contribute to future product planning process
   Not Important 1 2 3 4 5 Very Important   Poor 1 2 3 4 5 Excellent
3. New technology platforms are available for new product applications
   Not Important 1 2 3 4 5 Very Important   Poor 1 2 3 4 5 Excellent
4. Core research introduces new product ideas
   Not Important 1 2 3 4 5 Very Important   Poor 1 2 3 4 5 Excellent
5. Serve as a valuable knowledge base for new technology
   Not Important 1 2 3 4 5 Very Important   Poor 1 2 3 4 5 Excellent
6. Skill set of development engineers match future product development needs
   Not Important 1 2 3 4 5 Very Important   Poor 1 2 3 4 5 Excellent
7. Responsive to incorporating new features into ongoing development work
   Not Important 1 2 3 4 5 Very Important   Poor 1 2 3 4 5 Excellent
8. Bring new products to market quickly
   Not Important 1 2 3 4 5 Very Important   Poor 1 2 3 4 5 Excellent
9. Able to meet cost targets for new products
   Not Important 1 2 3 4 5 Very Important   Poor 1 2 3 4 5 Excellent
10. Cost of development is low
   Not Important 1 2 3 4 5 Very Important   Poor 1 2 3 4 5 Excellent
11. Contribute to the early concept phases of new product ideas
   Not Important 1 2 3 4 5 Very Important   Poor 1 2 3 4 5 Excellent
12. Technical skill level matches expectations
   Not Important 1 2 3 4 5 Very Important   Poor 1 2 3 4 5 Excellent
13. Communication channels are open and frequent
   Not Important 1 2 3 4 5 Very Important   Poor 1 2 3 4 5 Excellent
14. Time-to-market capabilities best-in-class
   Not Important 1 2 3 4 5 Very Important   Poor 1 2 3 4 5 Excellent
15. Other (please specify items not covered in survey that are important to you)
   Not Important 1 2 3 4 5 Very Important   Poor 1 2 3 4 5 Excellent

Product association ____________________

Salary Level ____________________

FIGURE 9.2 *Sample R&D Customer Survey for Product Management*

R&D Customer Survey for Manufacturing Engineering—Example

Directions: Rate each area in terms of importance to you as a customer of R&D, and your evaluation of our performance in meeting your needs in each area.

1. Paper or electronic documentation is complete
   Not Important 1 2 3 4 5 Very Important Poor 1 2 3 4 5 Excellent
2. Manufacturing participates on the development team
   Not Important 1 2 3 4 5 Very Important Poor 1 2 3 4 5 Excellent
3. R&D is a valuable information source for problem solving
   Not Important 1 2 3 4 5 Very Important Poor 1 2 3 4 5 Excellent
4. R&D supports cost reduction activities of the factory
   Not Important 1 2 3 4 5 Very Important Poor 1 2 3 4 5 Excellent
5. Product designs are optimized for factory tools and processes
   Not Important 1 2 3 4 5 Very Important Poor 1 2 3 4 5 Excellent
6. New product designs consider long term manufacturing technology directions
   Not Important 1 2 3 4 5 Very Important Poor 1 2 3 4 5 Excellent
7. Concurrent Engineering processes are used with manufacturing during development
   Not Important 1 2 3 4 5 Very Important Poor 1 2 3 4 5 Excellent
8. R&D provides product information to manufacturing early in the development process
   Not Important 1 2 3 4 5 Very Important Poor 1 2 3 4 5 Excellent
9. R&D is easy to contact and is responsive to requests for assistance
   Not Important 1 2 3 4 5 Very Important Poor 1 2 3 4 5 Excellent
10. The change rate (churn) of current products in production is low
    Not Important 1 2 3 4 5 Very Important Poor 1 2 3 4 5 Excellent
11. The change rate of new products being introduced to manufacturing is low
    Not Important 1 2 3 4 5 Very Important Poor 1 2 3 4 5 Excellent
12. Documentation is free of defects or errors
    Not Important 1 2 3 4 5 Very Important Poor 1 2 3 4 5 Excellent
13. The documentation format meets the needs for the documentation users
    Not Important 1 2 3 4 5 Very Important Poor 1 2 3 4 5 Excellent
14. Schedule commitments are met
    Not Important 1 2 3 4 5 Very Important Poor 1 2 3 4 5 Excellent
15. Documentation content and format minimizes redundant information entry and reformatting
    Not Important 1 2 3 4 5 Very Important Poor 1 2 3 4 5 Excellent

Product association ____________________

Salary Level ____________________

FIGURE 9.3 *Sample R&D Customer Survey for Manufacturing*

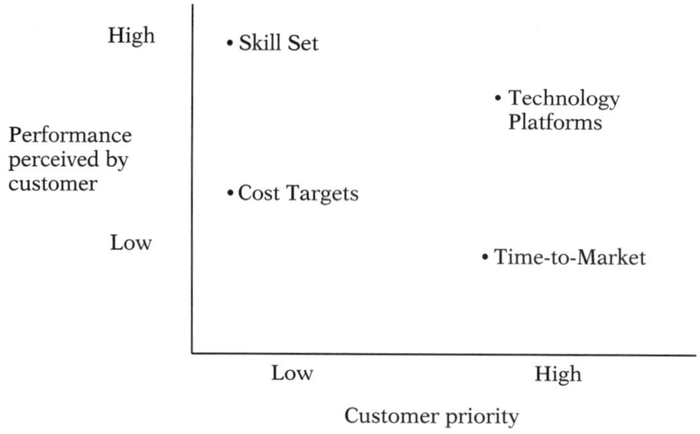

FIGURE 9.4 *Chart for Presenting Customer-Survey Results: Example for Product Management Survey—Telephone Sets*

generated. For example, you could have a table for working level associates, and one for managers. Moreover, you could separate the data by product type, as is shown in Tables 9.3 and 9.4.

What is important is that once a sufficient set of tables is created, common themes should begin to appear. These themes will be important input to the R&D directions that are to be established. For example, time-to-market appears as a common theme between telephone sets and telephone control units in Tables 9.3 and 9.4. In addition to time-to-market being a common theme, the performance is also poor.

A simple chart can also be used show the data from one of these tables. For example, Figure 9.4 shows example survey data plotted in a graphical way that simplifies the presentation of the material in Table 9.3.[2] A 2 x 2 grid could also be used to divide the graph in Figure 9.4 into four quadrants, placing each survey area into its respective quadrant of high performance–low priority; low performance–low priority; high performance–high priority; and low performance–high priority. This method of data presentation buckets the areas and helps facilitate discussion of your survey findings with others in your organization.

2. The data in Figure 9.4 and Tables 9.3 and 9.4 are for illustrative purposes only and are not based on actual organization results.

PERFORMANCE LEVEL AGAINST THE COMPETITION

Given a customer needs list and your customers' evaluation of your performance level, the only remaining piece of data is your performance level against your competitors and best-in-class companies. This data is collected in a process called *benchmarking*.

Benchmarks reflect best-in-class performance and serve as reference points for assessing current performance levels and setting objectives. Consider the customer need area of time-to-market. If your R&D organization can go from concept to ready-to-manufacture in twelve months for a certain product, and the *best* of your competitors can develop the same product in eight months, then the competitive benchmark for time-to-market is eight months. A noncompetitor, however, may produce a similar product in only seven months. This performance level would be the best-in-class benchmark.

The benchmarking process actually has two primary purposes. First, benchmarking establishes benchmarks. Second, benchmarking helps identify best practices employed by the best-in-class companies to achieve those benchmarks. In a sense, benchmarking allows you to adapt high-performance processes from other businesses to your operations (recall the discussion in Chapter 5 on process re-engineering). What is of interest in this section is establishing benchmarks for the improvement areas identified from the customer need information. To prepare for benchmarking, you need to identify the following:

1. What—the areas you want to investigate
2. How—the method for gathering information
3. Who—the companies to benchmark with

What—The Areas to Investigate

This step was partially covered in the first section when the customer needs were identified. However, what is needed to do benchmarking successfully is more than understanding the general areas. You must break down these areas into topics that can be focus points for benchmarking discussions. We conducted benchmarking with Sun Microsystems, Motorola, Hewlett-Packard, Florida Power and Light, Milliken, and other industry leading firms, and each time we prepared ourselves with specific questions and background material on the topics that interested us. We would spend three to five times more time preparing than actually being on-site with a company.

This preparation included researching the company in detail, brainstorming the list of questions on each topic area, sending selected material to the company in advance so that they could be prepared to deal with the topics, and last but not least, answering our questions about ourselves. Why? Because benchmarking is an exchange, not just a one-way transfer of information. When you are working with another firm, they will want to know as much about you as you want to know about them. You should be prepared to accommodate these needs, even to go so far as to ask for their specific areas of interest before the visit.

One mistake we made when conducting a site visit was not to share with the prospective company the details of what we were interested in, and *who* in their organization would most likely be able to have a meaningful exchange with us. In one case, we visited a company to talk about development processes and ended up talking to a sales person and a marketing person (those people who normally have contact with customers or other people outside of the company). The many days of preparation were wasted, and scheduling another visit turned out to be difficult.

How—The Method for Gathering Information

This step is most often thought of as visits to a company. Face-to-face visits are a valuable information source, but should only be done after all other public information sources have been researched. In the case of product competitors, it is unlikely that you would be welcome for a visit anyway. What is often overlooked is the abundance of information available in the public domain. These sources include, but are not limited to:

- Government publications and records
- Professional journals and publications
- Trade shows and publications
- Seminars and conferences
- University studies

Who—The Companies to Benchmark With

This step is completed by listing the primary "product" and "process" competitors to your business. Product competitors are those companies who compete directly with your products in the marketplace. Process competitors are industry leaders or other internal organizations with functions or processes similar to yours.

In some cases a benchmark from a direct competitor may not be as useful as a benchmark from a process competitor. For example, with telephones, the industry would be the electronics field. The lessons learned from a leader in the electronics industry may be easier to access than information from a direct product competitor. An example of this would be where a personal computer company can develop and manufacture a printed circuit board in eight months, compared to twelve months for a telephone manufacturer. The benchmark from the PC company may serve as a better reference point than your product competitor, and the practices employed by the PC company may be of more interest as well. In most cases, it is best to have both data points, so both product and process competitors should be identified on your benchmarking list of comparative companies.

Consultants can also be engaged to develop benchmarks. Most consultants have worked with many companies, possibly even your competitors. Although they cannot name a specific company for legal reasons, they often can make available to you the benchmarks for the market leaders in your industry.

Should you elect to do benchmarking on your own, the material presented here is not adequate to prepare you for effective benchmarking. We recommend the following authors who have detailed the steps necessary for successful benchmarking: Robert C. Camp (Camp 1989, 75–118), Michael J. Spendolini (Spendolini 1992, 105–183), and Gregory H. Watson (Watson 1992, 51–76).

After you have completed the benchmarking investigation, you need to assess the difference between your current performance level and the benchmarks of best-in-class companies. This gap analysis will be instrumental in goal setting, as described later in this chapter. This step is nothing more than saying, "We can do it in *X*, and the benchmark is *Y*." The difference between the two is the performance gap.

Then you must anticipate how these benchmarks will change with time. Understanding the rate of change and trends of benchmarks is critical. Overlooking where the competition and the industry are going can result in goals that leave a substantial performance gap when the goal is attained at some future date. For example, take a company with a time-to-market of twelve months that identified as a benchmark a competitor with a time-to-market of eight months. If the goal for this company was a time-to-market of eight months based on this current benchmark, and one year later the competitor achieved a time-to-market of six months, then the company failed to achieve a competitive advantage. This potential mistake is shown graphically in Figure

9.5. Note that the best-in-class benchmark was changing. As a result, the performance gap closed slightly over time, but the company seeking improvement did not gain a competitive advantage.

On the other hand, consider a company that believed they were outperforming the competition in a key area, but failed to continue improving in that area. The result could be a false sense of security now, and a substantial gap in performance at some future date (see Figure 9.6).

The end result of your benchmarking activity will be an established benchmark for each customer need area, and a forecast for how this benchmark might change in the future.

BUSINESS OBJECTIVES

The overall business directions must be considered when setting goals for R&D. Most companies have high-level directions and objectives for their business. To meet these objectives, the business requires that each organization develop goals that align with and support these directions. If the company is using policy deployment at this higher level, then the task is simplified for R&D. If, on the other hand, the business as a whole does not have clearly communicated directions and strategies, then the process becomes very difficult. People in product development may need to be the catalyst for changing how the organization establishes high-level goals for the business.

If the business has established high-level goals and directions, R&D has a responsibility to take actions that align with and support these directions. That means that, as an organization, R&D must understand how they can impact each of these areas. This can be done by breaking the business objectives down into specific areas that can be influenced by R&D. For example, take the business objective: *increase value of products and services for customers.* If you break down this area into its sub-components, the result would include items like those shown in Figure 9.7.

In this simple example, there are several areas that R&D does not impact directly, including billing accuracy and training availability. Other areas are influenced by R&D, including performance and quality. For this example, R&D would have to look at each of these areas to determine their current performance levels, and select an area to work that would contribute most to achieving that specific business objective. Another example would be the business objective of *globalizing* their products. A specific contribution that R&D can make to this objective is to gain competency in the environmental and type-acceptance regulations for each targeted country.

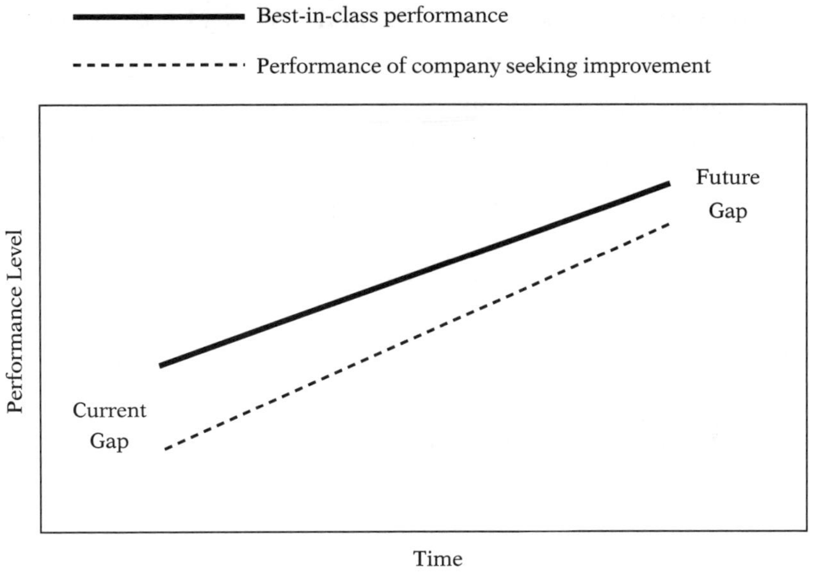

FIGURE 9.5 *Anticipating the Change in Best-in-Class Performance*

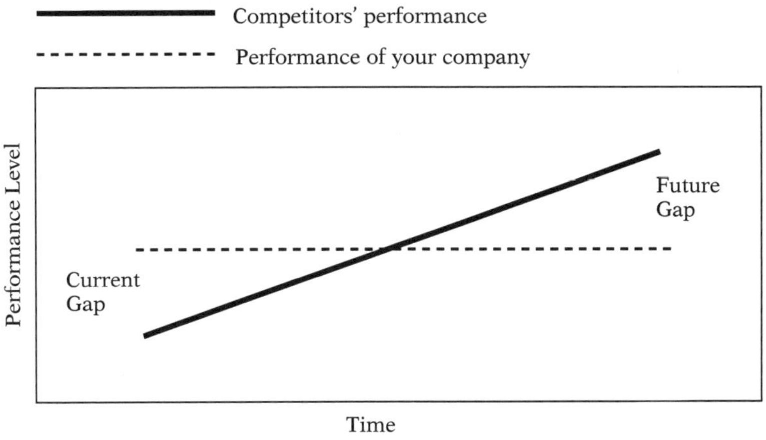

FIGURE 9.6 *Current Performance Gap May Give False Sense of Security*

While executing this process for each business direction, many areas for improvement may surface. You may want to interview your business leaders to prioritize these improvement areas for R&D. They can share with you how they believe the R&D community could best align with and support the business directions. This insight will not only help you with the process of goal setting, but it will also let them know that you are taking an active part in supporting the directions of the business, and what you are doing to set goals for R&D.

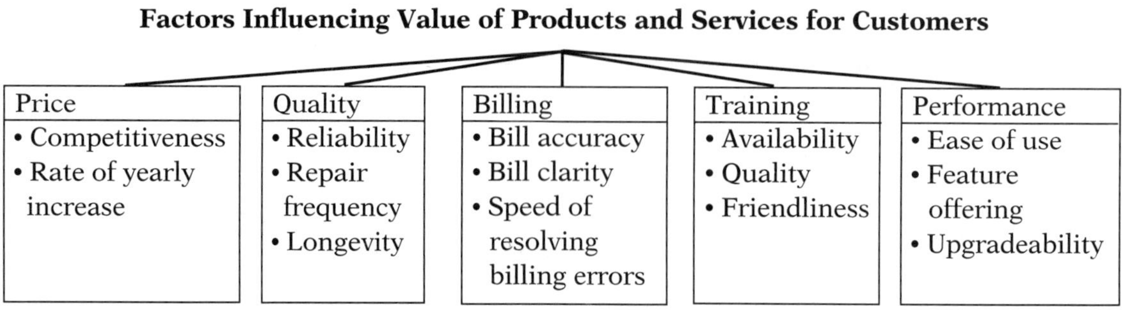

FIGURE 9.7 *Elements of Customer Value*

THE VISION FOR PRODUCT DEVELOPMENT

Visions are developed and shared to communicate clearly to the doers in R&D the future state of the organization. They can then integrate that vision into their daily work. Successful integration of R&D's vision into daily work allows each *individual* decision and action to push the organization that much closer to its desired future. The goals established in the goal-setting process should support moving in the direction of the vision for R&D.

Having a vision that people can understand and rally behind is an important part of the Quality Ideal, *Clear Vision and Goals Communicated to All.* The topics we have covered in this chapter are necessary inputs not only for the goal-setting process, but for creating a good vision as well. James Belasco, in his book *Teaching the Elephant to Dance*, details three criteria for good vision statements (Belasco 1990, 99):

1. Focuses on strategic advantages
2. Inspires and empowers; creates value for others
3. Is clear enough to be a decision-making criterion

To achieve the first and second criteria, you must have a strong focus on your customers and competition, as well as on the business direction and strategies. The elements covered earlier in this chapter are essential inputs to the vision creation process.

Our business unit in AT&T, Global Business Communication Systems (GBCS), serves the communication needs of small and large businesses. Our business unit's vision is:

To be your partner of choice . . .
dedicated to quality, committed to your success

This vision is very carefully crafted to cover three elements critical to customers. First, *partner of choice* suggests that GBCS will do what it takes to be the communication partner of small and large businesses. Second, *dedicated to quality* recognizes businesses' long-term view of GBCS as a supplier of the highest-quality products anywhere in the world. Third, *committed to your success* communicates to customers that GBCS is successful only when the customer is successful. Note that these same statements could be used for R&D, and address the needs of R&D customers as well. In other words, R&D could adopt this business vision as their vision for R&D.

It is important to note that the vision does not have to convey every positive attribute the organization would like to have. Visions are only a piece of a larger strategy that include the goals for R&D, the values (statements that reflect the beliefs and guiding principles of the organization), and plans, both strategic and tactical. Building too much into a vision makes it more of a motherhood statement than anything focusing and inspirational. The longer the vision gets, the less people pay attention to it, let alone remember it. Separating the vision from these other business elements allows the vision to be just that: a vision.

When helping your management establish a vision for R&D, make sure that they have a good understanding of the key customer needs and the current performance levels relative to your competitors. They should already be aware of the primary business directions, and should be encouraged to factor these strategic directions into their vision.

Before a vision statement is formulated, the vision for the business itself should be considered. When a vision for the business exists, an additional vision for R&D may not even be necessary. In other cases, the vision for R&D may be more focused than the vision for the business. When a vision for R&D is developed in addition to a general company vision, alignment between these visions should be very clear to employees.

An example of a good vision is taken from one of our R&D development groups. A hardware development organization located in Middletown, New Jersey, designs and introduces telephone products into a manufacturing organization located in Denver, Colorado. In 1985, both groups viewed one another as adversaries, each blaming the other for missed commitments, iteration after iteration of the design process, and errors in the product and documentation. The relationship was not exactly friendly. Over-the-wall

FIGURE 9.8 *Logo of "One Cycle, One Team"*

design transfer was the normal practice for transferring design data, and manufacturing was considered a second-tier organization by the development community (recall the story in Chapter 7 on concurrent engineering). With the introduction of CE in 1990, these two groups began a transition from their old, independent functional relationship, to an integrated, team partnership. These two organizations adopted the vision of "**One Cycle, One Team.**" The logo for this vision was a tandem bicycle, with two riders (the team) riding one cycle (see Figure 9.8).

"One Cycle" represented a single design iteration or design *cycle* that included all of the manufacturability requirements. "One Team" captured the singularity of purpose of the design and manufacturing efforts needed to achieve a high-quality design in only one iteration. Concurrent engineering, empowered cross-functional work teams, DFM, re-location of engineers, and other concepts were employed to make this vision come true. By 1992, the distinction between the organizations was only on the organization charts.

Note that the vision in this example meets Belasco's criteria discussed earlier. This vision focuses on *strategic advantages* in that reducing design cycles reduces the overall time-to-market for their product. It *adds value* for both the manufacturing and design people by partnering these two groups as a single team, working toward a common objective. Improved working relationships and equalized status of designers and factory engineers makes everyone's contribution recognized and valued. Finally, it serves as a *decision-making criterion* by steering all future process improvements in a direction that supports a design and manufacturing partnership and a single design cycle.

TABLE 9.5 *Goals for Product Development*

Business Objectives	Product Development Objectives
Increase customer value	Improve product reliability
Improve employee satisfaction	Improve employee satisfaction
Increase global growth	Reduce time-to-market for new products
	Increase level of forward-looking work
	Globalize new products

ESTABLISHING GOALS FOR PRODUCT DEVELOPMENT

At some point in time, all of this work with customers and competitors, visions, and business objectives needs to be brought together in a set of specific, measurable goals for R&D. This process does not lend itself well to a cookbook method. The sorting and coalescing of these inputs needs to be done in an iterative process involving many people and levels in the organization.

During this process, you will need to check the set of customer needs against the business objectives—i.e., do they line up? Recall that many of your customers are internal customers. Lack of alignment between these internal customers may suggest that the business directions have not been well deployed throughout the entire organization. If this is the case, some action needs to be taken to communicate to your customers the business objectives that R&D is using to develop their strategy, and to get them in line with the overall business directions.

One way to select the directions and goals for R&D is to lock the leader in a room and wait for a decision. An alternate and preferred method is to socialize all of the data with different levels of the organization and encourage dialogue within your own community. When it is all said and done, though, someone needs to make a decision. The result of this process should be a set of items that R&D will focus on, and the list should be short. An example list is shown in Table 9.5.

METRICS

For each area, you need to develop a set of metrics and measurable objectives based on competitor information and best-in-class benchmarks. For example, under the area of product reliability, metrics might include:

- Number of product returns (return rate)

- Percent of product dead-on-arrival (DOA)
- Number of calls to customer service center per day
- Average number of product repairs per month

One or two metrics that best measures the performance of R&D in this area should be selected. For each established metric, the current performance level of the organization should be documented, and short- and long-term objectives should be set. The time frames for the objectives may be in months or even years. Typically annual objectives are desirable, as well as metrics that project out three to five years. These objectives are based on the benchmarking information discussed previously in this chapter.

Once this activity is complete, your work up to here can be compiled into a simple-to-understand chart that shows the relationship of the R&D directions with the overall business, and illustrates the objectives for each direction. The alignment created between product development objectives and the business objectives is the first segment of your *golden thread*, a term we have used to describe the alignment of cascaded objectives down through an organization. An example is shown in Table 9.6.

Note that for each objective for R&D, metrics were selected to measure performance toward this objective. In addition, the current level of performance is shown. The one- and three-year objectives and targets are the result of comparing the current performance levels against the competitor benchmarks. The three-year targets should be set in such a way that they give you a competitive advantage in the marketplace.

Developing this table is an important milestone for this chapter. This table, if completed properly, reflects the directions and goals of R&D based on customer and business needs, competitor benchmarks, and the vision for R&D, all on one sheet of paper. Add to this sheet the vision for R&D, and you have all that is necessary to deploy the goals of R&D to the entire organization.

DEPLOY THESE DIRECTIONS AND GOALS THROUGHOUT THE ORGANIZATION

Communicating your directions and objectives throughout the organization is as important as creating them. In fact, failure to deploy these goals defeats the purpose of establishing them in the first place.

TABLE 9.1 *Sample Goal Table for Product Development*

Business Objectives	Product Development Objectives	Metrics	Current Level	1 Year Objective	3 Year Target (World-Class)
Increase customer value	Improve product reliability	• Return rate (# / month / 10,000 installed systems)	10	8	5
		• Customer trouble reports (# / month / 10,000 installed systems)	40	35	20
Improve employee satisfaction	Improve employee satisfaction	• Employee satisfaction index (% very satisfied)	55%	60%	75%
Increase global growth	Reduce time-to-market for new products	• Time from concept definition to market availability (months)	24 months	18 months	12 months
	Increase level of forward-looking work	• Percent of R&D budget allocated to forward-looking work	5%	8%	12%
	Globalize new products	• % of targeted countries where products are "qualified" and ready to be sold	20%	40%	95%

Depending upon the size of the organization, part of the deployment process is completed by the process used to create the goal table. If the R&D community is small, than the fraction involved in the creation of the directions would turn out to be large. The deployment process is mostly done. For large organizations, a more complicated process may be required to deploy goals.

The purpose of goal deployment is to align the activities of every individual in the organization with the directions established for R&D and the business. The objectives of departments and individuals should align with and support these goals. This action builds upon the golden thread discussed previously by aligning all of the plans and objectives throughout the organization so that they add up to the overall desired result for the company.

The recommended process for doing this is to first have an all-hands meeting (everyone attending), where the entire community hears, from the product development leaders, how the process worked, and what the results were. This ensures that everyone hears the same message, and any questions that surface are heard by everyone. The importance of having strong, enthusiastic leadership present the directions cannot be underestimated. People follow not only because someone has prepared a way to go, but also because they want to follow the leader. This is one of the places that the Organizational Enabler Leadership and Commitment at the Top plays a vital role.

The next step is for each part of the organization to identify plans and lower-level objectives that support these directions and goals. This process will require much negotiation and patience as every level in the organization must be engaged in this planning and objective-setting process. Many of the seven new management tools discussed in Chapter 6, including Affinity Diagrams and Interrelationship Digraphs, will be valuable in establishing plans and lower-level objectives.

Periodic reviews during the course of the year, as well as annual reviews, should be held to assess progress toward the objectives. If progress is not evident, then this gives managers an opportunity to re-allocate resources and develop new plans to ensure that the R&D objectives are met. This cyclic process is the plan-do-check-act element of policy deployment, and a critical part of actually realizing the goals for product development.

Part III

DEPLOYING QUALITY PROGRAMS AND PRACTICE

10 Developing Your Program

Strategic Quality Ideals

Organizational Enablers

✓ As you address the work described in this chapter, you will want to consider *all* the Strategic Quality Ideals (they're listed on page 19), and you will need to make use of *all* the Organizational Enablers (see page 19 to refresh your mind on these).

IT'S YOUR TURN

Up to now, the thrust of this book has been *educational*. In Part I we introduced the book, and put it into context. Part II was essentially a tutorial: we described what we have found to be important elements of any quality system or program of process management and improvement in a development or R&D environment. We discussed concepts, fundamental principles, methods, and tools. The tools we described ranged from various types of teams through tools for presenting data in ways that enhance understanding.

Now in Part III we are changing direction dramatically. Now that you have the basic tool kit you need, we are going to lead you through *implementing* and *carrying out* the things necessary for you to bring your own processes under managed control, and to begin improving them.

The first step, however, is for you to *develop your program*.

You need to answer the questions, "What do I want my organization to be like when we have achieved our goals? What sorts of changes does that mean our organization needs to make? What am *I* going to do to begin to bring this about?"

Why is it so important that you begin with this step?

If you are indeed right now at the very start of trying to change your development organization to focus on quality and on processes, you most likely have few (or no) supporters in your quest, and minimal resources committed to do anything. This is probably true whether you are the product development director for your company or the leader of one development team in a larger organization. From all we have experienced and heard, when some one person first catches the fever of what great things can be accomplished by a shift of focus to quality and to process, that person at first stands alone. This risky loneliness is not really new. More than four hundred years ago the Italian political philosopher Machiavelli wrote, "It must be remembered that there is nothing more difficult to plan, more uncertain of success, nor more dangerous to manage than the creation of a new order of things. For the initiator has the enmity of all who would profit by the preservation of the old institutions, and merely lukewarm defenders in those who would gain by the new ones" (Machiavelli 1947, 15).

One of the first things you need to do is to begin to spread the sense that this proposed change is a means to make the organization better: to improve market share, or increase customer satisfaction, or increase associate satisfaction, or whatever need for improvement you wish to satisfy. If you are at the top of your organization, it is easier to make some of the first steps happen—but you still face the problem of convincing others that you are on the right track, and that it is worth committing time and resources to the effort.

It is an *effort*. We are talking about a true paradigm shift, which can never be accomplished easily.

We believe that there are two important first steps:

- You need to have your own picture of where you want the organization to be at some future time.
- You need to build a case to help you get the support and re-sources needed to bring this about.

In order to accomplish this second step, you will need to understand where you are today, and then determine the changes that will be needed to move from your present state to your desired state.

Be advised that you are starting on a pretty difficult journey. In most cases, developing a quality system where there isn't one is not at all straightforward; it is a somewhat weird iterative process. You cannot get resources to make the changes committed until you can demonstrate the need and/or value of making the changes, but you need resources to get the data you need to put together the demonstration . . .

You will very likely need to start small. We discuss what this means in the section "Gaining Support," near the end of this chapter.

DECIDING WHERE YOU WANT TO BE

Experts in managing change (see, for example, LaMarsh 1994) say that before you begin to effect a major change, it is critically important to have a clear picture of where you want to be when you have accomplished the change. If you fail to do this, you will not only not know where you are headed, you will also not be able to tell when you have gotten there.

Your early view of where you want to be will necessarily be neither complete nor final. It will change and become refined as you change. But you need to start.

Develop a *personal vision*. In Chapters 3 and 4 we talked about the value of having a *vision*, and long-term *goals*. This is the point where you want to at least begin developing these, in terms of a personal vision of what you want your organization to be like. We say "begin" because you undoubtedly will want to involve your entire organization in fully developing and refining your vision and developing your goals.

The scope of your personal vision will vary greatly, depending upon what sort of organization you are trying to change. If you are working the change for an entire company, large or small, you will be looking at a more general future state, which will have to include the needs of your external customers. If, on the other hand, you are getting ready to improve one part of a development and manufacturing enterprise, your personal vision will be more limited, and focused principally on internal customers.

In order to describe your desired future state, think of organizations you would like to emulate. Do you know something about how your principal competitors perform—like how long it takes them to develop a new product, or something about their products that seems to be better than the way your product provides the same function? Back in Chapter 2, we offered a Five-Minute Self-Assessment. Do these questions suggest characteristics of your desired future state? For example (using selected questions from the Assessment), you might choose

- We know who our customers are, both internal and external.
- We have clearly defined processes for our work.

These were created by re-wording two of the questions in the Assessment into descriptive statements. Here are some other possible characteristics of the sort of quality organization you may wish to be:

- We have a clear vision and mission for our community.
- R&D goals and plans are well communicated and drive decision making.
- Functional and product-realization processes are well managed and improved *continuously.*
- Customer feedback and competitive information drive changes and improvement efforts.
- Day-to-day management decisions are based upon data, not on hearsay and guestimates.
- Associate satisfaction is measured, and feedback drives improvement opportunities.

You will want to add more items that represent your particular situation, and what you know about your customers and competitors. *Managing Transitions,* by William Bridges, (Bridges 1991) is an excellent reference that may help you to round out your vision.

You might want to list these characteristics of your desired future state down the left side of a page, with room for two more columns—as shown in Figure 10.1.

This brings us to the question of where you are today.

Desired State	Are We There?	Is Action Needed?

FIGURE 10.1 *Chart for Listing Characteristics of a Desired Future State*

UNDERSTANDING WHERE YOU ARE

You need to know where you are today for several reasons. The first, which we already talked about in the last section, is so that you can determine what you need to change to reach your target state.

Are the changes you want to make big or relatively small? Your answers to this question will have a major part in your deciding exactly how you want to go about changing: a real "re-engineering" project, or a program of continuous improvement.

There are a couple of other reasons you want fully to understand your present state relative to your target future state—so that:

- You realize that you *are* using processes today.

  Everything you do uses some process or other. You may not be aware of the process you are using at any particular time, and if you are not aware of it you certainly are not managing it.
- You can celebrate the knowledge that you are probably already doing some things pretty well—pretty much the way you want to be doing them when you have completed your program of changes.

  We engineers are notably poor in celebrating our achievements—we keep seeking ways to do even better. It feels good to realize that you are doing some things "right" already. Besides, this also means you have less changing to do.

Our experience is that this simple-sounding starting task, of identifying where you are today, is difficult to complete. The work of describing where you are is not functionally difficult; it is maintaining the discipline to *complete the task* that is difficult. Again we are battling classic engineer behavior: we want to get on with solving the problem! The tedious business of describing our current state is too easily seen as a distraction from the real task of changing things.

We encourage you to ignore this urge to "get on with it," and begin by doing a thorough job of describing your starting state.

Begin with the various characteristics you decided that you wanted to be part of your desired future state. For each one of them, ask yourself, Do we do this now? or, Do we look like this now? Write down your answer in the form of a statement—like this, for example:

We do not have clearly defined processes, that are described in writing.

Be completely honest; there is no value in "cheating" even a little. At this point only you are aware of what you are writing, and knowing where you *really* are is critical to success in moving to a new state.

Here are some other questions you might ask yourself. There is some overlap with our previous suggestions, but the different form in which these questions are asked may trigger you to write down some more characteristics of your present state.

- As you do your development work, are people aware of the processes they use, or are they focused almost entirely on getting their projects done, by whatever means—occasionally even "heroic ones"—are required?
- Are you aware of and managing your overall product realization process (PRP), or is there tension and a sense of competition between marketing and development (and manufacturing)?
- Are your development intervals competitive, or are you constantly losing sales to your competition because you are late getting new products to market? Does marketing keep threatening to go outside with their next development project?
- Are your costs to develop and manufacture new products competitive, or are your sales people unable to set prices at a level that allows an adequate profit margin because the basic product cost is too high?
- Does the factory have trouble manufacturing "good" product (which passes all tests with plenty of margin) because your designs are not sufficiently robust to tolerate being made with the normal variations in parts and components?
- Are you taking advantage of the new trends in using teams and empowering your people, or are you still working the old way, with close control by managers at all levels? Are you "pushing decision making to the lowest possible level," or do one or more levels of managers still have to approve everything?

Checking Against a Reference Point

To help you to understand exactly what you are doing, you may want to get a handle on your present level of performance, perhaps by comparing with some "standard." You do not want to do this in great detail at this point—there will be opportunity and more resources to do that later. You might pick one of several criteria against which to check. Perhaps the general definitions from the Software Process Maturity Framework, from the Software Engineering Institute at Carnegie Mellon University (Humphrey 1989, 445) will help you place yourself. The SEI suggests classifying software processes into five levels:

5:	Optimizing	Data are automatically gathered, defect-cause analysis is rigorous, and improvements are fed back into the process.
4:	Managed	("Quantitative"): A measured process, with at least a minimum set of Quality and productivity measurements in place.
3:	Defined	("Qualitative"): The process is defined and institutionalized.
2:	Repeatable	("Intuitive"): The process is dependent upon individuals; there are basic project controls but there is no orderly framework for improvement.
1:	Initial	("Ad hoc" or "Chaotic"): There are no formal processes, and no management mechanism to ensure that even the informal processes are followed; senior management does not understand the key issues.

If your work is more hardware than software, Computervision has developed a similar maturity model called the *MMEI Maturity Level* for manufacturing and mechanical engineering design.[1] Another quick check you can make is to compare yourself with the criteria for the Malcolm Baldrige Quality Award (refer back to Chapter 2). Remember that winners have tended to score above 750 points.

PLANNING HOW TO PROCEED

Identifying the Gaps

Now that you have identified where you want to be, and where you are today, it is time to identify the gaps.

1. For information about conducting a Mechanical Process Assessment following this system contact D. H. Brown Associates, Inc., 222 Grace Church Street, Port Chester, NY 10573.

Take out the chart you started, per Figure 10.1, and fill in the second column, "Are We There?", based upon what you have determined about your present state.

Now complete the third column: "Is Action Needed?"

How much action is needed? Are you "out" on just a few characteristics, or on most of them?

For some of the "Action Needed" characteristics, you may be able to go a step further, and at least estimate the distance you are away from where you want to be. This will, of course, depend upon the type of information you have. If you rated yourself against the Baldrige criteria, you may have a rough numerical gap. If you know your own present development cycle time, and have an idea of how much you would like to improve this, you have an idea of your outage. You can write this sort of information into another table, like the one shown in Figure 10.2.

Look at the "Change Needed" column. How big are your problems? How big are the changes you need to make? If you can, make an estimate of how difficult it will be to bring about this change.

You will probably want to keep this chart where you can refer to it often, and update it as you learn more.

In Chapter 5 (Figure 5.1) we introduced the *Process Health Matrix*, shown again for easy reference in Figure 10.3. The Process Health Matrix is a tool that can help you translate your understanding of the magnitude of the change that lies ahead into a plan for action.

Dimension	Present State	Desired State	Change Needed	Change Difficulty
Interval	24 months	12 months	50%	

FIGURE 10.2 *Table for Estimating Action Needed*

Process Management	Process Performance: Poor	Process Performance: Acceptable and improving
Effectively implemented	Use problem-solving tools (Chapter 6) 1	2 Continue with process management; focus on metrics and benchmarking (Chapter 4)
Not present	3 Implement re-engineering (Chapter 5)	4 Implement process management (Chapter 4)

FIGURE 10.3 *Process Health Matrix*

To enter the health matrix, you will need to separate your "Action Needed" characteristics into two groups:

- Those that describe whether or not you are today using concepts of process management
- Those that characterize your Process Performance today

For instance,

- If you have clearly defined, documented processes, that is an example of Process Management that is Effectively Implemented.
- If your software development process rates Level 1 on the SEI criteria, and your goal is to be at Level 3, that is an example of Process Performance which is Poor.

Make this evaluation for all of your "Action Needed" Characteristics. Is your overall process management rating "Effectively Implemented" or "Not Present"? Is your overall process performance rating "Acceptable and Improving" or "Poor"? Based upon these two evaluations, in which quadrant of the Health Matrix do you place yourself?

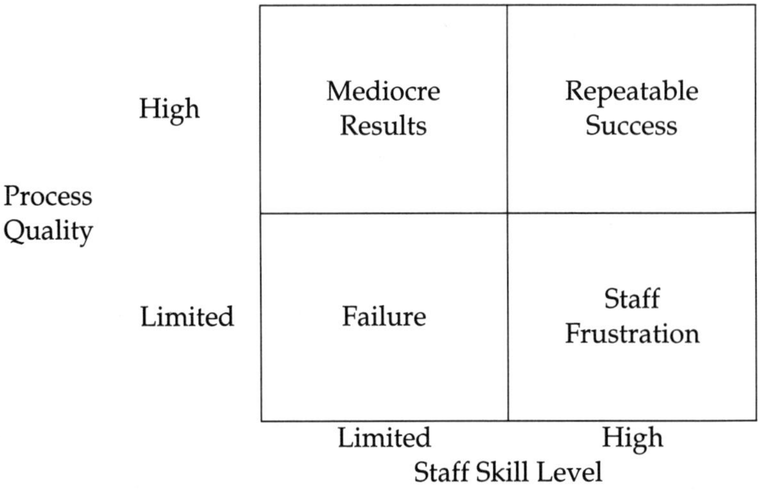

FIGURE 10.4 *Process Quality and Staff Skills Interact to Support Successful Development of Quality Products*

Depending upon which quadrant you find yourself in, go back and read the appropriate sections in Chapters 5 through 7.

Another potential gap is the availability of the needed skills. To achieve the requisite high levels of quality requires a skilled staff. How prepared are the people in your organization to do the jobs that will be required of them? Assuming that they all have the appropriate formal degrees, are they up-to-date in their fields—especially in the new areas that will characterize your work in the next few years, and the latest methodologies and tools? They may need some professional development to close this gap. All will need some quality and process training; these are discussed later in this volume.

The matrix of Figure 10.4, which is based upon one used by AT&T's Group Technical Officer for the Network Systems Group, graphically puts into perspective the roles of quality processes and staff skills in consistently developing quality products (Stanzione 1992). This figure says that quality processes can to some extent overcome limited staff skills, but that truly *repeatable* success can be achieved only by providing a highly skilled staff with quality processes to use.

You will probably want to build into your program the needed training experiences for your staff.

Drawing Up a Starting Plan

You now should be ready to write down a basic starting plan. You won't be ready yet to write down the *contents* of the plan—that will come mostly after you have gathered a team and resources and really started your activity to change the way your organization works.

What you can write down now is an outline for your plan. Doing this will help you to consolidate in your mind the task that lies ahead of you. Having the outline will help you describe your proposed activity to others whose support and participation you seek.

Your plan will of course be focused specifically on your organization and the needs for change that you have identified. What we present here is a general outline that contains items we believe to be important for almost any change to a quality and process-focused environment.

A. Leadership for R&D
    1. Vision and mission
    2. Long-term strategic plans and goals
    3. Annual objectives
    4. Goal-deployment mechanism
    5. Review and feedback process
B. Process Management
    1. Customer satisfaction data
    2. Competitive performance assessment
    3. Process documentation, including process descriptions
    4. Training
        a. Functional
        b. Process and project management
    5. Process Metrics
    6. Do we want our quality system to conform to ISO 9001?
C. Associate Involvement
    1. Team structure
        a. For product development
        b. For process management
    2. Should we use self-directed teams?
    3. Training
    4. Performance evaluation and reward system

You can use this outline as a starting point. Keep those items that you believe you will need, drop others, and add items that you have identified as necessary to achieving the end state you see for your organization.

At this point, this is as far as you need to go.

The next step, which you may actually want to begin developing as you work with key people to gain support, is to specify for each item when you need it and how you plan to attain it—including the people who are key to attaining it. For instance, for the first item, vision and mission, you will probably want to identify:

- When will we need to have these available, to tell everyone?
- Who will actually draft the vision and mission?
- How will we get them to do this job? (For instance, should they go on a multiple-day, off-site retreat?)
- Who will be asked to comment on the drafts, and how will we gather these comments?
- How will we promulgate the finished vision and mission to everyone in the organization? For instance, should we use policy de-ployment?

The final step of filling in the answers can come only after you achieve a starting level of support, and begin to assemble your team. We will therefore pick this up again in the next chapter.

Gaining Support

You need to devise a way to persuade the people who are key to your obtaining the needed resources that what you want to do is important for the organization. Chapter 11 goes into detail on lining up resources and building the support structure; what we are talking about here is the basic spadework you need before you can start the work described there. You need to get the people who hold the resources to listen to you. We believe that there are three basic ways to achieve this:

- Scare them into believing that if you do not make the changes you propose, the organization (perhaps the whole company) is at risk.
- Run a pilot or trial activity, on a very limited scale, that will produce some data to show the potential benefits of the changes you propose.
- Convince your boss by some other means, regardless of what level you are, that the state to which you want to move the organization is good for the organization and for the whole company. You may be able to do this by exposing your boss to external information—through published data, a

speaker, a conference, or perhaps a class—demonstrating that this is the right way to go.

If you can directly convince your boss of the value of your proposals, then your whole job will be easier. You may, of course, have to do the convincing via one of the first two approaches listed above.

If "scaring" the organization into supporting your ideas seems like the way to go, you might try something like

- Conduct a small customer-satisfaction study—external or internal—to find out how your organization's customers feel about your performance. Do you meet their needs? With some room to spare, or marginally?
- Conduct a survey of the associates in your organization, to learn
  - if they are concerned about the company's performance relative to customer needs or the competition, and
  - what they know about the areas represented by the characteristics where your desired future state is different from your present state.
- Interview senior managers, to see if they believe the company (organization) is competitive.

Even if you do not seem to be too far from the performance level of your competition, remember this: the competition is also running—maybe even faster than you are!

If you decide that the best way to go is to run some sort of trial or pilot to get data with which to convince others of the value of what you propose, you might start by trying to enlist the people working on some project that has just begun or is about to start to do some things in a way that matches some characteristics of your desired future state. You do not need to totally revamp how they work. You can probably get them to agree to try a couple of things without taking on too much extra work, or jeopardizing their budget or schedule.

If you have stayed with us this far, you have the outline of a plan for moving your organization to a focus on quality and process, with some notes on when each element needs to be completed, what resources you anticipate needing to accomplish each element, and what approaches you might employ to gain support. In the next chapter we address how to gather the needed resources and build the support structure you will need to carry out your plan.

11 Lining Up Resources and Building the Support Structure

Strategic Quality Ideals

✓ Total Employee Involvement and Empowerment

Organizational Enablers

✓ Leadership and Commitment at the Top
✓ Support and Alignment among Middle Managers
✓ Strong Values Shared by All
✓ Reward Structure Linked to Contribution

Now that you have an idea of what you want to achieve, and how you want to go about it, you need to get resources (the funds and the key people) committed so you can begin your work. You also need to put in place a support structure so that you can apply these resources most effectively.

In this chapter, we look first at *Obtaining Funding*. This has two important steps:

- Prove the need to your funders so that they commit to fund the work
- Get a budget approved so that you actually have funds to spend

With funds assured, you can then begin *Gathering Your Team* of key players—the people who will work closely with you to begin implementing your plans. While you probably have already had some help as you developed your plans, this is the time when people can begin to devote substantial amounts of time to quality work, rather than working only "in the margins."

To support these first participants in your effort, there need to be certain elements of *The Support Structure* in place. We conclude the chapter by looking at

- Architecting a structure, so that people see where they fit
- Providing common language and tools, so that the people can work together more easily
- Beginning to build enthusiasm

This last step is important as you begin to engage your entire organization; it may be even more important to help your small team of key people get past the difficult early days when they seem to be alone in the quest to change to a quality-focused organization.

Keep in mind as you undertake this step that the single most important element is gaining and keeping the unwavering support of the top manager or leader of the organization that needs to change.

An important corollary to this is that if you are the agent of change, focus on the part of the organization that you lead or that you have the support to change. Success in this area will help you later to sell the changes in other parts of the organization.

OBTAINING FUNDING

Prove the Need to Your Funders

The first step is to prove to your funders (most likely your top managers) that they need to support your planned changes to an environment focused on quality and process, and also support the work needed to implement the changes. This is a more focused effort than was described in Chapter 10, with more specific goals. The goal in Chapter 10 was to get their attention. By now, you should have more data and a better idea of your plan; your goal now is to obtain *commitment* for resources sufficient to begin to achieve your planned changes. Three steps you can take to gain the support of your funders are:

- Make sure that your funders are aware of the importance of quality in all aspects of the business, and of process management as a means to improve quality
- Show your funders data that you have gathered, which show what performance *can* be achieved
- Show your funders data from your own operations to show where you are and point out opportunities for improvement

This is a natural continuation from the tactic we called "scaring them" in Chapter 10. Let's consider each of the steps in turn.

The Importance of Quality

You must make sure that your funders are aware of the importance of quality in all aspects of the business. In the 1990s, American industries, professional societies, and universities have begun to appreciate the importance of having the highest-quality products, services, and processes. This learning is increasingly available in a variety of formats (including published material, conferences, and symposia) that you can use to convey your message. Here are some approaches:

Provide Written Material. Chapter 3 identifies a number of journal articles and books as references. Because their work is so well known and highly regarded, references by or about Taguchi or Deming or Juran are especially forceful. (See, for example, Taguchi 1993; Scherkenbach 1986; Dobyns and Crawford-Mason 1991; and Hunt 1992.)

Members of major professional societies have access to the Engineering Societies Library, in case references like these are not available locally. The Library is located at the United Engineering Center in New York, which is also the home of the ASME, IEEE, and other professional engineering societies.

Show Videos. Have a showing at work or let your funders take the videos home to watch. Many of the same organizations that sponsor training programs also rent or sell videos, which range from packaged courses (the SME does a lot of this) through lectures by experts such as Dr. Deming.

Provide Training Experiences. Conferences and symposia dealing with quality and process management and concurrent engineering are sponsored by professional societies and private enterprises. Some of these sources are given in the box provided here. Attending a conference is one of the best ways for people to get up to speed quickly, and at the same time perhaps to begin personal networking with colleagues in other companies who have similar interests.

Representative Sponsors of Quality and Process Conferences

- The American Society of Mechanical Engineers, 345 East 47th Street, New York, NY 10017
- The Society of Manufacturing Engineers, One SME Drive, Dearborn, MI 48121
- The Institute of Electrical and Electronic Engineers, United Engineering Center, 345 East 47th Street, New York, NY 10017
- DFM Alert/The Management Roundtable, 1050 Commonwealth Avenue, Boston, MA 02215
- American Productivity and Quality Center, 123 N. Post Oak Lane, Suite 300, Houston, TX 77024-7797
- Institute for Competitive Design (formerly Troy Conferences), 134 W. University Drive, Suite 108, Rochester, MI 48063

Some specific topics that you may want to be sure your funders become aware of are:

The Role of Process Management and Improvement

The same experiences that so clearly demonstrate the importance of having high-quality products and services also demonstrate the importance of quality processes. Process management and improvement is a critical route to achieving the desired levels of process quality—and hence quality products and services. Consider again the concept that "Fixing the problem fixes it this time; fixing the process that caused the problem keeps the problem from recurring in the future."

The Role of Skilled Staff

To achieve the requisite high levels of quality also requires a skilled staff. Both the company's hiring processes and training programs affect this. The hiring process, which is outside the scope of this book, must identify and bring on to the staff people with the types of skills appropriate to the jobs to be done, at the appropriate start-up skill levels. This is true for both recent graduates and experienced "market" hires, who, when carefully selected, bring ready-to-use critical experience. As mentioned in the preceding chapter, both groups need ongoing training in their basic skill areas, in the latest methodologies and tools, and in process- and quality-related topics.

As pointed out in Chapter 10, quality processes can to some extent overcome shortcomings in staff skills, but truly repeatable success can only be achieved by a highly skilled staff.

Staff frustration can hurt achievement in several ways—not the least of which in good economic times is that the frustrated high-quality staff will seek more personally rewarding opportunities elsewhere, leaving you with only the less-skilled staff members.

Show How Much Can be Achieved

You can show your funders what can be achieved through a focus on quality and process in a number of ways, including sharing with them stories of others' successes or the results of benchmarking activities.

Quality Improvement Success Stories

Sharing stories of the success of others is a way to convince through the power of authority and universality.

There are documented success stories from companies that turned their businesses around, recovered lost market share, significantly improved product quality, or achieved major improvements in the time to introduce a new product. Some sources are in the References section of this book. Where we refer to certain journals or magazines, check other issues of the same publications. Other books, journals, and conference proceedings available from the resources listed in the box earlier in this chapter may yield examples directly relevant to your business. Here's an example from AT&T that, while it is not from product development, demonstrates very clear results.

Example. Certain AT&T sales offices were having trouble meeting delivery and installation commitments to customers who had purchased particular office communications systems. When the office staff members looked for the root causes, they recognized that the problems were in their support processes. A Quality Improvement Project (QIP) was set up to examine the provisioning (order processing, delivery, installation, etc.) process for these communications systems, and the related customer service and support processes. Their solution was to create the role of *project manager* to oversee the entire systems provisioning process. This allows the account executive (the sales person) to move on to another sale as soon as the contract is finalized on one sale. The project manager meets with all concerned parties, including the customer, to establish the schedule for installing the system, conducting training, etc. Measured results of this process improvement included

- Customer satisfaction increased from 41 percent to 88 percent, "excellent"
- Customer willingness to repurchase increased from 53 percent to 82 percent

As you gather quality and process stories, look for companies who make similar products, or at least use similar technologies, to those of your own company. Focus on *reported results*: stories of true successes are the most convincing. Note what quality programs they used to obtain their results (see Table 3.1 for examples).

In his books *In Search of Excellence* (Peters and Waterman 1982) and *A Passion for Excellence* (Peters and Austin 1986), Tom Peters[1] presents several stories of companies that have successfully applied quality and process management

1. Tom Peters is a well-known business author, whose *In Search of Excellence* (Peters and Waterman 1982) helped to trigger interest in the United States in improving corporate productivity.

concepts to achieve dramatic results. Hewlett-Packard and John Deere are other companies that have successfully applied these concepts.

The Xerox turnaround story is well known. The authors first heard David Kearns, Xerox CEO, describe it in the keynote address at the AT&T Bell Laboratories Quality Day in 1988. H. Barry Bebb, who had been Xerox vice president for manufacturing during the early turnaround, described it at an NSF-sponsored Conference of Design Theory & Methodology at the University of Massachusetts at Amherst in 1989, and again in the keynote address of the ASME National Design Engineering Conference in 1991. In 1992, Kearns and David Nadler published the story (Kearns and Nadler 1992). Briefly, Xerox was surprised by a sudden loss of market share to Japanese copier manufacturers. They benchmarked the competition, making use of their alliance with Fuji Xerox in Japan to obtain the data. They found several measures, including quality in manufacturing, cost of engineering, and time to change manufacturing setups, that clearly showed why they were being bested by the Japanese competition. They realized that the loss of market share was nearly terminal. Starting at the very top with Chairman Kearns, they committed to making drastic changes in how they did business, and set metric goals for themselves that would enable them to catch up with and pass the competition. They attained their goals for change, regained much of their lost market share—and proceeded to set new, tougher goals for *continuous improvement*.

Cadillac's use of multi-function teams to improve the quality and rapidity of their product developments has been widely publicized since they won the Malcolm Baldrige Award (McCusker 1992). Florida Power & Lights' wide use of empowered teams helped them to win the Deming Prize in 1989.

Thinking of power companies, here is a story, undocumented and very likely apocryphal, that points out the importance of pressing to find the root causes of problems. This story was heard by one of the authors at a training session (SME 1992):

> *There had been frequent outages along one section of highway somewhere in the United States, due to power poles being knocked down. Since the problems occurred in an area known to experience an unusual frequency of lightning strikes, it was "concluded" that the cause was lightning. It was decided that the best way to deal with this "act of God" was to station a special crew in the area to effect quick restoration of service whenever one of these lightning-caused outages occurred. After the crew had been in the area for a while, they reported back to headquarters that the outages were indeed storm related, but not caused by lightning strikes. It seemed that there was a sharp curve in the*

road, and when the road was wet an unusual number of skidding accidents occurred—many of which took out power poles, which were located on the outside of the curve, and thus caused power outages. The new solution, which solved the problem, was to move the poles to the inside of the curve.

A second aspect of this story was cited as an example of the value of involving everyone in the organization in the continuous improvement effort: the members of the special team who were sent to deal with the "lightning problem" eliminated the need for their special assignment when they reported back the real root cause. The person relating the story speculated that had the team not felt that they were fully involved (i.e., were they still behaving the "old way"), they might have chosen not to alert management to the real problem—in order to prolong their independent "special assignment."

Another power company story is about a series of problems attributed at first to lightning (an easy culprit to blame, since when it does occur it can cause major problems) that turned out to be due to migrating birds perching on power line towers (Gabor 1990, 175–176). The root cause may not always be as dramatic as the first identified "cause," but it usually is amenable to solution!

You may find that one or more of the stories from AT&T that we have presented are relevant to the plans you want to put in place, and can help you show your funders what can be achieved through the planned activities.

Similar success stories from within your own company should carry even more weight with your funders. Such stories often appear in company newspapers or technical magazines, and are described in detail at annual *Quality Day* or *Process Improvement Sharing Rally* gatherings. A phone call to the author or presenter will normally produce a detailed memorandum describing the successful activity; people who have achieved such successes are usually anxious to share their learning with others in the company, to help the company do even better as their ideas spread.

The company's quality organization is another potential source of success stories. An advantage here is that you can talk with the folks who had the experience.

Benchmarking Data

Benchmarking is another tool for demonstrating the importance of carrying on process management and improvement work. The basic idea is to compare how certain performance measures in your own company compare with the same measures from other companies. These data are strongest if they come from your competitors.

You started preparing to do this if you undertook the assessment recommended in Chapter 2. While you can not undertake any major benchmarking activities until you have funding, you may be able to find published data, or to obtain information from an organization like the American Productivity and Quality Center, whose address can be found in the box of sponsors earlier in this chapter. They hold benchmarking seminars and discuss benchmarking results in their publications. You need to seek both the internal data and external data from other divisions of your own company, from competitors, and from noncompetitors who make "similar" products. What sort of data are most useful? Consider, as examples:

Intervals. What are representative intervals? Be sure projects you are comparing are similar in magnitude and have similarly-defined starting and ending points. Combine the best sub-processes from several similar processes to define the "best-of-the-best" interval. Figure 11.1 shows an example. From benchmarking, you are able to determine that Company C has the shortest planning process, Company A the shortest product realization process, and Company B the shortest development process. If you could emulate these three best sub-processes, line D shows the hypothetical best-of-the-best product realization process that you could possibly attain.

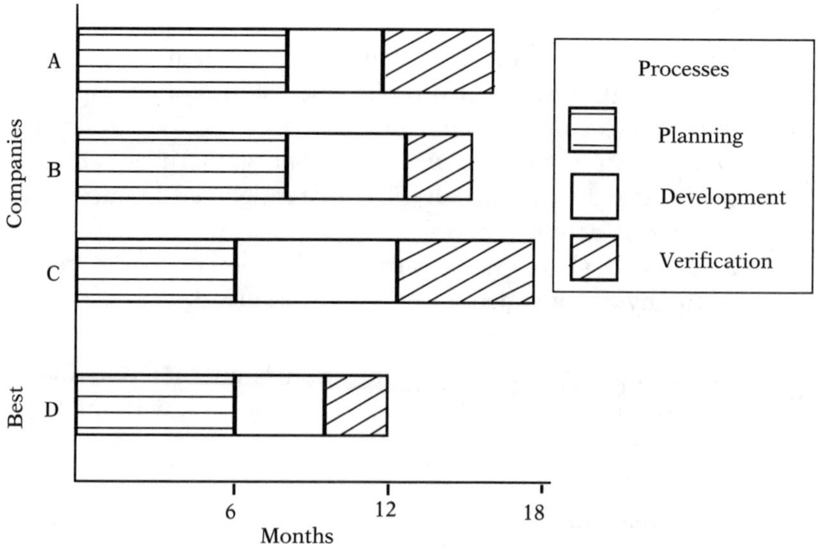

FIGURE 11.1 *A Single, Composite Product Realization Process, Comprising the Three Best Sub-Processes, Is Shorter than Any of the PRPs from Which It Is Developed.*

Repair Information. If your own company has other manufacturing divisions, they may be the easiest place to find information on returns under warranty or for other repairs. What is the factory's return experience? What are warranty experiences? Do products returned for "repair" often show no symptoms of trouble when the repair people go to work on them?

Show Data From Your Own Operations

Unless your company was founded just recently, you have plenty of data that you can use to show opportunities for improvement. Your first inclination may be to state firmly that you have no data, since you have "not yet begun to collect any." But it is very likely that you *do* have data—lots of it. You need only to find the raw data and present it in usable form.

If you did some sort of pilot project as suggested in Chapter 10, you should by now have data from that.

Next, look at the history of your most recent projects.

Intervals

Look at records of the last few projects that were completed. When did they start? When were they released for manufacture? How much time and money were spent in drafting?

Group these most recently completed projects by size and scope. Look at overall time to complete development, and at the intervals (durations) for major parts of the development activity, such as

- Settling on the definition of the product
- Hardware development
- Software development
- System test
- Field evaluation
- Preparation for volume manufacture

Now look at projects that are underway or being scheduled. What do they predict for the same intervals? How do these compare internally with your previous projects? And how do they compare with benchmark data?

Slip Ratios

One way to measure the effectiveness of the processes people use is to observe whether or not schedules are regularly met, and if not, by how much they are not met. "Slip ratios" are a way to factor out project size or scope from the schedule-slip records of several projects, and so get a better feel for trends in process effectiveness.

For each project, if the original schedule was not met, divide the total amount by which the project completion date slipped past its original schedule by the originally scheduled PRP interval. This is the slip ratio for that project. Compare these for all of your recent projects. Did they get better? Did they get worse? It is likely that they showed no particular pattern—which is a good argument for devoting effort and funds to process management and improvement.

Design Transfer

Look at your record of transmitting design documents from the development organization to the factory, especially the changes and re-transmissions. Why was it necessary to re-send the information? Don't stop with just the apparent reason; seek the *root causes.*

Figure 11.2 is one list of possible causes, from the authors' experience. The one to really look closely at is "OOPS", which includes things like typographic errors, incomplete information, mistakes in the design, and just plain goofs. If one is honest about assigning changes to the OOPS category, there will probably turn out to be an embarrassing number of changes with this root cause.

> **Example.** When a Quality Improvement Project (QIP) led our engineers to assign root causes to all of their design documentation changes, and they saw how many they were forced to admit were OOPS, they began to seek ways to do it right the first time. The group that collected these data on 160 design transmittals found that more than 38 percent were in the OOPS category. This example was discussed in more detail in Chapter 6; Figure 6.2 is a *fishbone diagram* showing what problems they identified. After their recommended process improvements had been in place for about two years, the OOPS rate had dropped to about 10 percent. Thus the present level of OOPS design documentation transfers is an excellent indicator of an opportunity for improvement.

Another category of documentation changes that may well indicate an opportunity for process improvement is "Requirements changed." These probably extended your product realization interval in several ways, starting with a de-

Reason for Change	Change Type
Failed external criterion test (FCC, UL, CSA, etc.)	OOPS
Failed internal safety requirement	OOPS
Stocklist error	OOPS
Drawing error	OOPS
Different component value needed	Design Error
Requirements changed	Life cycle
Cost avoidance	Life cycle
Cost reduction	Life cycle
DFM improvements	Life cycle
Design for test improvements	Life cycle
Re-rate product (e.g., no longer to be manufactured)	Life cycle

FIGURE 11.2 *An Example List of the Root Causes for Engineering Design Changes Transmitted to Manufacturing*

layed start and certainly including re-designs if they came in the middle of the development work.

Design Robustness

Records that show the robustness of recent designs may provide another set of opportunities to shorten the development interval through process improvements. Did you pass UL tests the first time, or was rework needed? If your product line includes electronics, did you pass FCC Part 15 Emissions tests the first time? How much margin does the product have? Is the margin so narrow that the factory will have a hard time manufacturing units that pass required periodic retests, because component values are within specification but on the other end of the tolerance band from the components in the unit that passed the original FCC tests? A more robust design, with more margin, will pass the retest. Is some close-tolerance mechanical part hard to make consistently? The control charts and yield data may point to an-other opportunity to improve productivity through process improve-ment focusing on robust design.

Establish a Budget for Process

Let's assume that your efforts have paid off, and you have convinced your funders to invest company funds in your planned changes. The next question you can expect will be something like How much do you need, and exactly how do you plan to spend it? You need to prepare a *budget* for your program.

The first step is to identify for what exactly you will need funds. Look at your plan. For what specifically do you need money? To start up an effort to change an organization to be focused on Quality and on process management and improvement, regardless of the details of your plan, you will need funds for:

- A core team of *change agents* (you and your immediate co-workers)
- Training
- New tools to support improved processes
- Salaries for people who can be expected to be drawn into the new work as your plans are implemented

Later on you may want to have funds allocated to permit everyone to charge some of their time to process improvement, but this will not happen during the first year of your work.

The Core Team

Since you are working on this budget, the core team of change agents probably has already been launched: you and your team. From the authors' experience, there probably need to be *at least* three people on this core team, so that you act as your own support group, generate more ideas, and so that you can work with various parts of organization at the same time. When estimating the funding needed to support this core team, do not forget:

- The team's managers/coaches, who will be involved and will surely charge at least some of their time to the process work
- Clerical support for the core team, which by taking care of routine matters will give the team more time to work the professional issues
- Some expense money, to cover things like
  - Travel to conferences and for benchmarking visits
  - Computer software to help the team members do the new kinds of tasks they will be undertaking
  - Books and journals, so that the team members can keep pace with this rapidly changing field
  - Various items to support the culture changes (these are discussed in a later section; the important thing at this point is to provide some money for them)

Training

What sorts of training will need to be funded? While, as already mentioned, it is vitally important that all members of your product development team are fully trained in their technical specialties, it is assumed that training to support this goal is not funded as part of the process management and improvement effort. You *will* probably want and/or have to fund training programs that directly support your quality improvement efforts.

What sort of training might this include? At least the following:

- Training in the philosophy of quality and process work—the sorts of topics discussed in Chapter 3. This training would most likely be in the form of formal classes, taught by consultants or perhaps obtained from some satellite vendor like NTU (The National Technological University)
- Training to establish alignment within the company, by deploying your vision, mission, and high-level goals
- Training in your chosen methodology for process management and improvement. Everyone needs to understand this methodology and be able to apply it.
- Training in using problem-solving tools
- Team-building activities

Who should receive what sorts of training? Managers need to understand the concepts and philosophies of quality and process management, and need focused training in establishing and nourishing alignment of everyone with the vision, mission, goals, values, etc. The core team members need more specific quality and process management training; they probably will spend more time in process training than anyone else—for them this is also their functional training. They will also use the skills they learn to train others in quality and process management skills.

It is also absolutely necessary that *everyone else* in the organization receives quality and process training. A truly successful program requires that *everyone* understands

- the philosophy you are using,
- the basic methodology you are using, and
- tools appropriate to their responsibility level,

so that they can routinely apply them to their work. How much quality and process training should you consider budgeting for everyone? If your company requires forty hours of training each year for every employee, consider one day for this training. If you can convince your funders, add another day of *alignment* or *team-building* training for everyone.

How much needs to be budgeted for all this training? There are, of course, the wages of the trainees. There will also be class costs: fees if you bring in outside classes or even just facilitators, trainer wages if you use internal instructors, and costs for materials (e.g., texts or videotapes) if you use them.

Everyone Works on Quality

The final piece of your quality and process management budget has to be a means for *everyone* to charge some of their time to this work. The cleanest way is to provide special charging cases for quality. If there is not a specific charging case for quality work, experience clearly shows that many of the people simply will not spend time on it without a great deal of encouragement from top management. Eventually people will charge quality work to their regular project accounts; top management support is crucial to ensure that their project leaders don't at some point—certainly if money gets tight—tell the people and you, "not with my project money!"

How much budget should you seek for this? While in companies committed to improving quality and well along in their changes, people spend more time on process work, as your quality and process work is getting started you may want to seek management support for perhaps 5 percent of the total development budget.

Budget Approval

Getting the budget approved will be part of gaining management support. The problem is fairly simple: if they don't buy into the concepts, they won't approve funding. If the amount of money you need requires additional budget approvals, or if the organizational funding structure is such that development leadership needs parallel approvals, you may have more convincing to do. If you prepared your immediate funders (top managers) well, you can expect that they will be significant champions to help you obtain all the needed approvals. A good *business case*, fully detailing effort, resources, schedule, and expected results, should be a significant help.

A caution is necessary here: watch out for some non-believing "approver," perhaps in your product management or marketing organization, who thinks that the 5 percent set aside for process work is a place to save some money for a pet new project. To protect against this person's cry that "We don't need this process stuff," you will really need to have your development funders aligned with the goal of making the changes you have proposed.

GATHERING THE TEAM

Identify Key Players and Sell Them on the Need

The core team of process management and improvement "doers" or change agents was introduced in the budget discussion earlier in this chapter. For mutual support you probably want to have a minimum team of two persons in addition to yourself—even if in a small company one of them can only participate part-time. Other key people will be the core team's managers/ supporters, and the leaders, already in place or needed, for project- and function-based quality and process management efforts within the organization. The managers/supporters of these other leaders are also very important people, but before they can be approached the working leaders need to be identified.

Since it is very likely that all of this quality and process management activity is either brand new or at best unfamiliar in the organization, it is not possible to simply identify candidate doers by looking at the technical credentials or present assignments of the folks in the organization. Other approaches need to be employed. You certainly want people who are users of the processes that will be affected. "Since the people engaged in the process already know more about that process than anyone else, they will also know best what it takes to fix it. Grass-roots involvement, then, is critical to the success of any program aimed at changing the behavior of the people performing the process" (Patterson 1993, 23).

The *absolutely critical requirement* in the early days is that each person recruited to this work has to *want to do it*. Being a change agent and trying to change the corporate culture is difficult, often discouraging work. A person shanghaied or coerced into it will not have the enthusiasm to be a sufficiently powerful role model or the staying power to last through the discouraging times.

How, then, can you find the few people who have appropriate technical and human relations skills, and who want to be part of the "new fangled" work of improving quality and productivity in R&D?

Evangelizing

You yourself really need to be a believer if you are going successfully to lead the organizational and culture changes needed to make quality and process management the "normal" way of doing business. Really internalizing these concepts is almost like a religious conversion. This may sound hokey, but it's really true! You'll recognize it when it happens to you. Instead of continually challenging these new ways of approaching the job, you will clearly recognize their value. You will *want* to do things differently, and you will recognize that underlying everything you do is some sort of process. You will understand the idea that

> *By improving the Project, you fix the current problem, but by improving the Process used to do the Project, you improve not only the current Project but all future Projects.*

Apple Computer actually formally designated some key team members "Evangelists" (Kawasaki 1990, 99), recognizing that introducing some new concepts needs to be done in a new way, with a passion not familiar to many people in the business world. If you have this sort of job—and introducing the concepts, methods, and culture of quality and process management certainly is this sort of job—and you don't fully internalize it, you may not succeed. You may not be able to survive the hard and frustrating times you will experience: the resistance from nonbelievers, and the constant challenges from people who feel quite comfortable with the old ways. You certainly won't be able to sell others on the importance of the new ways.

As you begin to feel comfortable with the new ways yourself, you can start to identify other believers. There will be some other believers—maybe only very few at the beginning, but as the work progresses, more will appear. They may be trying on their own, within their own sphere of influence, to bring about the same sorts of changes. You may find a software team leader who has identified the process the team is using, and is working with the team to document the process and work improvements. You may find someone who is trying to convince the organization that the new European *ISO 9000* standards are important to the business, and that the organization should be working toward being certified in those standards. You may find someone who has just been to a conference where one of the acknowledged leaders of the new methods—perhaps Dr. Taguchi himself—has convinced your colleague that your organization needs to do things differently. Enlist these individuals as disciples. You may want a few who have a sufficiently broad view and are well respected throughout the organization to become part of your core team. Others will serve best by continuing to work in their functional or project organizations. Wherever they are located in the organization, network closely with these people so that they support your goals both by what they themselves do and by spreading the message of the new ways by word and by example.

Socializing

This is the more structured (conventional?) approach to selling the concepts, with which you can begin to reach the key people in the organization who are not as ready to embrace the new ways, but whose support is absolutely critical to the job. These people include the managers/supporters of the people who do believe and want to bring about the changes, upper management, and leaders of related activities that need to be integrated into the complete product realization process—leaders from marketing, manufacturing, etc.

Alignment

Building alignment within the organization—of vision, of mission, of objectives, of goals, of the importance of total quality and of process focus—is both a necessity and an excellent means of socializing the concepts on which alignment is developed. The need for alignment was discussed in Chapters 3 and 9. Alignment needs to extend vertically throughout the organization, from the CEO through the ranks of management to the least-skilled worker. A gap in vertical alignment will mean extra work for you at levels below the gap, convincing those people to follow the message from "upper management" (often defined as anyone two steps above the speaker) rather than that from their immediate management. The importance of tying in middle managers, who are often among the people least willing to change, was pointed out in Chapter 3.

There is also clear need for horizontal alignment with your colleagues. If you are to reach the workers in your own organization, you really need to have the support of their managers/supporters. Their vertical alignment upward, even if it is clear and strong, is most likely to be in fairly general terms—of things such as *vision* and *mission*. You need their support for the more detailed parts of your program, which is extremely difficult without some degree of alignment upon which to build. Your local management can help with this by including the principles of your program in their alignment message.

Expose People to What You Have Learned

Make sure that all of the key people in the organization have access to all of the good information you have accumulated about the quality movement nationally, and about the benefits of process management and improvement. Distribute to them the information from conferences and benchmark visits in which you have participated. This information can include literature you have picked up, plus substantial trip reports. Be sure to focus on the *data* you have accumulated.

You should also distribute other pertinent literature that you have found to be useful, including journal and magazine articles and books. Should you find what you believe to be a "super book," buy one for every manager in the organization, plus the key quality- and process-oriented people you have already identified.

Encourage People to Learn on their Own

When you circulate books and magazine articles, you hope to whet people's appetites, and start to build their interest to the point where they will seek more information on their own. Instituting a noon hour "video theater," where you bring a VCR into a conference room and show videos by people like Deming or Juran or Tom Peters, can be immediately instructive and build enthusiasm, and also may encourage the viewers to read more by the same people—especially the books on which the videos are based.

Circulating conference and short-course announcements may also encourage some people to participate, and thus continue their learning on their own.

You may want to invite speakers to give seminars to parts or all of your organization. While inside speakers are less expensive, outside speakers may have more influence. You may even want to follow the lead of the safety folks, who put up posters and encourage people with slogans.

THE SUPPORT STRUCTURE

Architect the Environment

It should be absolutely clear that you cannot create a culture focused on quality and on process management all by yourself. You also cannot achieve the desired ends by involving only management. *Everyone* must eventually be involved. To reach all of these people and draw them in is much easier if an environment exists that supports participation in the new culture and methods.

Create a Process Structure. Whether your present structure is *project* oriented, *functionally* oriented, or *matrixed*, you will want to superimpose upon it at least a minimum structure to carry out the tasks of organizing, documenting, managing, and improving your processes. The authors recommend a Process Management Team (PMT) structure, where for each identified process of significance a process management team is established. Each PMT is typically staffed by a few of the more enthusiastic quality and process management supporters from among the users of the particular process. They will look to you for guidance in what to do and how to do it, for which you will want to create a *process for process management;* how to do this was described in Chapter 4.

Plan to Involve Management. We have already said a great deal about involving management. Their involvement becomes an important part of the environment for change. From the environment point of view it is especially important to have top management involved and serving as role models. If they can *lead* in practicing the new ways, rather than *mandating* change, as was often done by top management even in the recent past, they will provide powerful support.

Plan to Take Advantage of Teams. The use of many different types of teams was discussed at length in Chapter 8. Among the first teams you may want to set up are process management teams to own and manage your key functional processes, such as software development or user documentation development. Are you likely to want a process management team to oversee the whole product realization process, and attend to optimizing the interfaces between functional processes? Will you want to sponsor training to help the entire organization buy into the concept of concurrent engineering teams—maybe even self-directed teams? Establishing key teams early and planning for other appropriate teams will clearly help you build the support structure you need to move toward improved quality.

Common Language And Tools

Especially because of the magnitude of the change required to move the engineering staff from a *project* focus to a *process* focus, it is very useful to bring everyone in the organization (or, at least those who will be involved first in the quality effort) together with a common vocabulary, and with common tools. This is especially important for the core team leading the process management and improvement effort.

Common Language

One important step is for your core team to take the basic training classes together. This provides some team building, and also ensures that everyone will share the same view of the new concepts and the same definitions of key terms—starting with *process management.* This not only ensures shared information within the team, but also ensures that when people throughout the organization work with the process team, they will see and hear consistent messages.

Common Tools

Process management and improvement utilize many new methods and tools that will probably be unfamiliar to the people in your organization. Specific methods and tools are discussed throughout Part II of this book, especially in Chapter 6, where you will find details on a number of tools that have worked well in the authors' organization.

Some tools with which you may already have some experience are

- Quality Function Deployment (QFD)
- Robust Design
- Statistical Design of Experiments

Describing these in detail is beyond the scope of this book; sources are listed in the References.[2]

From the tools we have described, you may find these most useful in the early stages of your work:

- Hoshin Planning (Chapter 4)
- Process Quality Management and Improvement (PQMI) (Chapter 4)
- Brainstorming (Chapter 6)
- Affinity Diagrams (Chapter 6)
- Interrelationship Digraphs (Chapter 6)

Training, which was discussed earlier in the context of planning time and funding for it, plays an important role in giving your process team a common tool set. You may be able to take standard training courses to meet your organization's needs from your own in-house education organization, from professional societies, from universities, or from consultants.

You may also feel a need for customized courses, tailored to your specific business and using examples from your current development projects. Many people seem to learn more quickly and more thoroughly when the examples are clearly relevant to their own work. We have successfully used courses where the processes examined in breakout group exercises were the actual processes being used by the members of the breakout groups. After the course is over, the group can continue to develop the work they began in class, to understand better and improve the processes they work with daily.

In-house education organizations can usually offer this sort of customization, as can some consultants. Once your core team has completed a module of training, you have the opportunity of making an even greater impact on the rest of the organization if there are process team members who can teach the

2. In the References, see Clausing and Hauser 1988 for QFD, Taguchi 1993 for Robust Design, and Christensen 1977 for Design of Experiments.

course. You can really customize the course, incorporating into it really relevant examples and supportive messages. In addition, your team's understanding of the material is significantly reinforced as the team members teach the material to their colleagues.

Cautionary Example. One thing to be alert for, as people begin to master the individual tools: can they put the tools together so as to obtain the best solution for a real problem? With engineers, you can expect some success with numeric tools. The "Seven New Tools" (see Chapter 6) are another matter! They are essentially nonnumeric, and are targeted to poorly-defined problems where one needs first to get a handle on what the problem really is. After our own organization had completed a trio of courses, covering the basic process management methodology and two sets of tools, we held another one-day class to address this very topic: how to use the tools together to solve real problems. We prepared a half dozen examples, "like real" problems, of increasing complexity, designed so that four or five tools, including some of the nonnumeric tools, could reasonably be applied in order to obtain a "solution"—in the case of our class, a direction for further action. We were totally unprepared for the class's performance! People who appeared to have a good grasp of the individual tools, and how to use them when told which tool to apply to a particular situation, were essentially unable to select the appropriate tools to apply to open-ended problems. We ended up spending more time that day lecturing than having the class carry out exercises, and we modified future tools-training experiences to focus more clearly on when to use which ones.

Quality Cascade

Another training tool that may prove useful, especially to disseminate the basic concepts of quality and of process management, is the *cascade*—where each level of management teaches each training module to their direct reports, cascading down through the organization until everyone has received the training. This combines the relevancy of customized training and the reinforcement value of teaching what you have learned, with alignment that is developed as each team is trained by their leaders. When the cascade is completed, everyone has heard the same message, and understands that the message is universally accepted throughout the organization.

It has been suggested that maintaining a consistent message in this sort of cascaded training is helped by having the prior-level trainer sit in on each training session. An example should make this clearer. Consider the four-level organization shown in Figure 11.3.

A trains the three *Bs*, who then train the *Cs*, who then train the *Ds*. If, when the *Bs* train the *Cs*, *A* sits in on each session, *A* can help ensure that the message is passed on without loss of any important content. This consistency really helps develop the alignment that is a major goal of this sort of training.

Begin to Build Enthusiasm

Because process work is important, it needs to be recognized and rewarded. Some of this reward needs to be part of annual compensation: built in as a necessary, focused part of performance review. People have to understand clearly that at least some part of their annual reward is based upon their contributions to process and quality. In addition, set some money aside to give special rewards to your best process doers and leaders.

Rewards

One area to which you will need to pay close attention is how your existing "traditional" reward structure supports your desired new environment. This was discussed in Chapter 8 with specific reference to teams. In many companies, especially in development, people have for years been hired and rewarded based upon their abilities as individual contributors. It is a shock to both the individuals and the system suddenly to change the focus to teams. You need to understand how existing performance evaluation processes need to be changed to recognize contributions to process and to quality clearly, and to recognize contributions by individuals as members of teams.

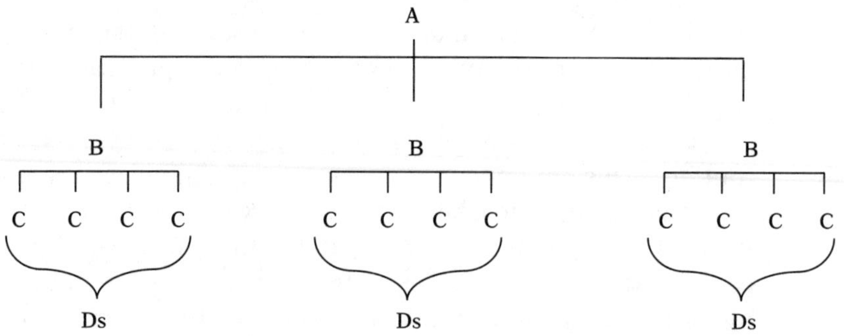

FIGURE 11.3 *Example of Cascaded Training*

Beyond Compensation

People work for recognition as well as monetary reward. One of the authors recalls from a number of years ago, reading salary surveys that said specifically that this is true for engineers, essentially all of whom have salaries that more than meet their basic needs. Because of this, you will also want other, special awards. This topic is discussed at length in Chapter 14.

You may also find it appropriate from time to time to provide tokens that promote the concepts of quality and process management and improvement not as rewards but as reminders. Among these might be

- T-shirts or polo shirts with logos that identify specific projects or teams
- Desk or wall plaques reiterating key company beliefs, like a vision or value set
- Pocket cards with the same sort of information on them

12 Getting Started

Strategic Quality Ideals
✓ Clear Vision and Goals Communicated to All
✓ Focus on Process; the How of Doing Work

Organizational Enablers
✓ Leadership and Commitment at the Top
✓ Support and Alignment among Middle Managers
✓ Strong Values Shared by All
✓ Reward Structure Linked to Contribution

RECAPITULATION

In the preceding chapters we have laid out the important steps for implementing Quality and process management programs, plus the methods and tools needed to implement those steps, and led you through some necessary preliminary stages. As a reminder, here are the key topics that have been covered:

- Establishing a vision and a mission
- Picking a process management methodology
- Choosing between incremental improvement and re-engineering
- Solving problems with methodology, tools, and data
- Using teams for maximum effect
- Setting goals
- Employing concurrent engineering
- Developing the program you want to implement
- Gathering the needed resources and support structure

Now, let's put it all together and get going.

THE PRELIMINARIES

How You Start Depends Upon Where the Impetus Is Coming From

The very first thing you need to do is note where the impetus to change to a process-focused organization is coming from. Is it an initiative begun at the very top of the company, or is it coming from the grass roots—perhaps even from you, yourself? Or is this a local management initiative, with support some distance up but not nearly at the top? Where in the structure are the people who approved your funding?

Depending upon the answer to this, your start-up work will need to emphasize different aspects. If the impetus for change is coming from the very top of the company hierarchy, you will probably receive high-level goals, along with a vision and a mission. You will not need to struggle nearly so hard to devise these key concepts, but you will need to develop aligned local views to the goals, vision, and mission that are passed down to you. The high-level support structure will probably be in place. You may even get cascaded down to you

- A process management methodology
- Tools
- Mandates to use
  - Teams
  - Concurrent engineering
- Suggestions about whether or not to re-engineer some element of your work

Your funding resources may have been identified with no effort having been made on your part to convince anyone, and you may even have been given some benchmark data that were gathered by the top executives during their training activities, that may well have included one or more visits to companies that have won the Baldrige Award or have otherwise been recognized for their accomplishments in moving to being quality-driven and process-focused.

At the other extreme, if yours is a grass-roots effort, you may want to start out by allying yourself with others in the company, in or near your organization, who have similar interests and desires. For instance, are there efforts in your organization to bring product development and manufacturing closer together? Are there people in other parts of your company (or in your own organization) who are promoting the potential of a process focus to improve development quality, intervals, and costs? Are there local champions for "Design for X" or DFM?

Then, either alone or with the compatriots you have found, you need to acquire a basic set of relevant skills by taking appropriate training. With the knowledge and confidence this will give you, and depending upon your level in the hierarchy, you will probably need to sell someone above you on supporting you until your process activity attains the character of a local management activity, at some level where at least a small program can be implemented.

In this case, you will need to do all of the "get ready" activities, at a level appropriate to the level and size of the supporting organization.

Fleshing Out Your Program

Now you can begin to flesh out the program you planned in Chapter 10, using the resources you obtained and the structure you set up following the approaches described in Chapter 11.

If Change Is Driven from the Top

If you are working with a mandate from the top, much will already be laid out for you. You will need to decide what you can do to support the goals and plans of the corporation. You must understand what you can achieve with the staff resources you have available, within the funding guidelines that are provided.

You will also need to set up the local environment. How will you convey to everyone the company's goals, vision, and mission? You will need a local support structure, which will have to be started by engaging the managers and establishing your core team of "experts". You will need to get this basic group trained in

- Using the chosen process management methodology
- Using concurrent engineering
- Using teams
- Using the mandated and recommended tools

If Change Has Begun Down in the Organization

On the other hand, if it is (or when it becomes) a local management activity, your first steps will be to probe upward for higher management support, and then get some basics in place, including:

- Goals, vision, and mission
- A local support structure—which requires that you engage those managers who will join in at this early stage, and establish your core team of experts. With luck, you can find a few real potential champions

Cautionary Example. Be prepared for some surprises during this time, depending upon what people have already heard about "this process stuff." When we were given the opportunity to expand our process work from our "pilot" department to the next larger development organization, we approached one person who we believed could make a really good contribution and who we thought had some understanding of and interest in making the change to a process focus. We were not prepared for the opposition we encountered: some others in the organization, who apparently mistrusted or feared the proposed changes, actually advised our candidate *not to join us*. Their advice was that the only truly valued work in the organization was work on projects, and that moving to process work would be destructive to the individual's career. If you should encounter something like this, it's best to move on to another candidate. Do not use up your energy fighting. Demonstrated results will eventually convert most of the nay-sayers; some small fraction may never buy in to the process-focus approach.

Now you can begin to set up the environment. In this case, though, you will have to *choose*

- A process management methodology
- Whether or not to use teams, and of what types
- Some tools you want to use

As soon as you have made these choices, you can proceed to get your basic group trained in

- Using the chosen methodology
- Using concurrent engineering
- Using teams
- Using the chosen tools

The next thing you will probably want to do is establish a benchmark baseline. Since you will probably want to do this fairly quickly so as not to lose momentum, you should try to use benchmark data that are already available to you. Perhaps you acquired some useful data as you were developing your program, or during your training. If you found folks elsewhere in the company who are ahead of you in this change, they may have data they can share with you. Depending upon your type of business, there may also be useful data in the literature.

Pick a Pilot Project

To guide you as you gather your first key people—managers and your core team of experts—it will help to have in mind where in the organization you will first implement your changes. Pick a first project, depending upon the job and the nature and size of the organization with which you are working. This may be the same pilot you used to help you to develop your program, back during Chapter 10, or it may be another project that seems more appropriate at this time.

Should you be lucky enough to be working with an organization that includes both development and manufacturing, which will give you the best opportunity to exploit the benefits of concurrent engineering, look for a fairly small, self-contained project that is in the concept development phase. In an electronic company, this might be the re-design of a major circuit pack. A new model of a successful consumer product would be another example, or perhaps converting some significant machine component from a weldment to a cast part. Such a project can use the new approaches through development and into standard manufacture, while involving a relatively small number of people. These scale limitations are important, because you want to be as sure as possible that you will succeed, and thus be able to demonstrate to the entire organization that this is the right way to go.

If you are working with an organization that includes only product development—without integral manufacturing—most of the same considerations apply, but you will want to select a group and a project that will be sufficiently well defined for you to be able to demonstrate the results. In an electronic company, you might still select a circuit pack re-design, because the interval from concept until a working circuit pack is demonstrated to pass critical tests (e.g., Underwriters Laboratories and FCC) is well defined, and can be compared with similar intervals in the past or in other groups not following the new approach. Converting a weldment to a casting would be less successful, because until manufacture gets under way the full benefits may not be clearly demonstrable.

The criteria for selecting a first project bear summarizing. They should comprise:

- A relatively small number of people, since it may be easier to convince them to try the new approach, and it will be easier for you and your small core team to support them as they work the new way
- A fairly small, well-defined, self-contained project that will come to a sufficiently clean completion for you to be able, in a reasonable time, to demonstrate the gains from working the new way

Share Your Goals and Vision

At about the same time you are choosing a pilot project on which to try the new ways of working, it is appropriate to share the goals and vision with everyone in the organization. (You may want here to refer back to Chapter 9.) If the goals and vision have come from higher in the company, this may have already been done—or it may be the right time to continue to deploy them. If they are locally generated, it is clearly the time to start. You want *everyone* to share the goals and vision, not just the few who are working with you and the people whose project will form the pilot. A key reason for getting everyone involved to at least this level right away is that your pilot will be working in the larger organizational context. While the pilot group may be able to maintain enthusiasm and discipline within their project, they will inevitably interface with other people in the organization. *It is critical that these others at least understand what is driving the people who are striving to work the new way.* If this is not assured, the pilot group may well be daunted by misunderstandings as they interact with others who are not involved in their trial of the new ways.

Quality Policy Deployment is a proven means for sharing your goals and vision with the rest of the organization. This method was described in Chapter 9; here is an opportunity to use it to achieve important results.

As you are deploying the goals and vision may be an appropriate time to begin building enthusiasm. Use the deployment sessions as kick-off meetings to show your own (and management's) commitment and to build enthusiasm. Maybe give everyone some sort of "reminder" item: a billfold card, or perhaps something to hang on their employee badge. Relatively inexpensive, convenient cards like these keep your basic message in front of everyone; the other side may display some related information like the basic steps in your chosen process improvement methodology, or the Shewhart cycle. Have the card made of light card stock, and plasticized for durability so that it stays clean and readable.

CONRAIL (The Consolidated Rail Corporation) goes a step farther, to make sure that their quality message is omnipresent. In addition to featuring their

"Continuous Quality Improvement" logo on employee business cards (as well as on locomotives!), they put on the back of business cards these basic quality messages:

> *We, the employees of Conrail, are dedicated to making our company the carrier of choice in every transportation market we serve.*
>
> *We promise safe, reliable and innovative services that meet or exceed customers' expectations.*
>
> *We are committed to continuous quality improvement as a means of providing superior service to our customers, developing and recognizing excellence in one another, enhancing value for our shareholders and being worthy of the public's trust.*[1]

Every time an employee handles or gives away a business card, the company's quality message is right there.

Make Sure of Your Basic Resources

It is time to check on all of the things you were urged to do in Chapter 11, to line up the necessary resources, and prepare a support structure.

Make sure that your basic resources are lined up. Given your chosen pilot project, have the necessary managers agreed to be part of the trial? Is your core team of process professionals (experts) set?

Have you proven the need to your funders, so that you have the needed financial resources? This probably will require that you have established a budget and gotten it approved (whatever that means in your organization).

If possible, have you identified other believers? They may prove useful if you encounter opposition along the way.

Understand How to Split Up the Work

You now should have a pretty good idea of the size of your start-up job. Depending upon whether the impetus for a process focus comes from the top of the corporation or from your local organization, you have an idea of the size of

1. Courtesy of Conrail.

the job you'll have to do convincing others to join in. You know what your pilot project will be, including how complex it is and how many people will be involved. You should have a good fix on your available resources: the money, and the people on your core team who will be selling and coaching the others.

With all of this knowledge, go back to the process architecture you outlined as you developed your program (remember Chapter 10!), and refine it to fit your actual activity. You may at this point be focusing only on part of it. Don't give up your longer-term goals, but do redraw the architecture to show people what pieces you will be working in the short term. Then use this to see how to split up the work of your own people. You want all of the parts you have chosen to work on to be staffed, you do not want people to be overloaded, and to the extent possible you want people to be doing the things they are best suited to do.

STARTING YOUR ACTUAL QUALITY WORK

Okay, it is now time to really get going. Keep in mind that a major part of your focus will be on the pilot project, and the people doing it.

Beyond that, you will probably want to stick to

- Other parts of your process architecture that you have chosen for short-term implementation
- Deploying the message of quality and process focus throughout the organization
- Helping management spread the word to peers, and upward if necessary, to make the environment as receptive as possible to the changes you are implementing

Establish Process Responsibility

Get the pilot project folks organized to work on process management and improvement along with their project work. This basically means that they need to establish

- Any team(s) they are going to use to carry out the project—self-directed or not

- The roles of the various coaches (managers) involved in both project and process work
- A process management team (PMT) to work the specific tasks of process management and improvement. The members should represent at least the major processes that will be used to achieve the pilot project. (This is elaborated below.)

This is a good time to refer back to Chapter 8, to see how the team concepts apply to your specific situation.

You will need to have identified members of your core team to be process coaches for each team that is formed for the project. You will probably want to work with the coaches in their new roles—they will need *lots* of help, especially if the project uses any self-directed teams.

Do you want to use one or more self-directed teams? This may be difficult to answer. The first times you employ self-directed teams, there are some difficulties and some risks. For this first pilot project, you may not want to make this big leap—especially if people are totally unused to teams, and appear to be concerned about forming teams at all. On the other hand, if the climate seems right for employing self-directed teams, *do it*! The potential great results are probably worth the risk and the extra up-front work (e.g., for training and for going through the "Forming, Storming, and Norming" period 3).

An outstanding tool for acquainting managers with the issues involved in implementing self-directed work teams, and suggesting how to get started making the necessary changes, is the book *The Empowering Supervisor*, by Susan S. Stahl (Stahl 1992). Written in the form of a novel, in the spirit of *The Goal* (Goldratt and Cox 1986), Stahl's book is enjoyable to read and conveys the essentials clearly and painlessly.

Understand Your Current Processes

This is the first job of the process management team, with the help of the rest of the people working on the pilot project. They will probably need lots of help from you and your core team as well.

At this point you may want to review the material in Chapter 4.

What ARE the Current Processes?

For the pilot project, what overall process will be used to accomplish the job? What functional processes will be involved as sub-processes to the overall process? Is it starting early enough that marketing or systems engineering processes are being used? What major development processes will be used? There may be circuit design, mechanical design, software design, system integration, and/or system test processes. What about manufacturing—what processes will be used to get the new product ready to send to customers? There may be manufacturing engineering, manufacturing process engineering, purchasing, material or storeroom management, packing, and shipping processes.

This is a good time to re-examine the membership of the PMT. Did you get all the right people? Is every major sub-process represented by an "owner" or leader?

Document the overall process and the major sub-processes, following the methodology you have chosen (refer back to Chapters 4 and 10 if necessary). Remember to begin with the *purpose* of each process. The software design process, for example, has a fairly clear purpose: to design and develop the software needed for the product to function. Identify who "owns" each process. The owner of a process should be an expert (preferably with *profound knowledge* in the sense used by Deming), who is respected by the peers who will use the process to do their work, and who is empowered to make improvements in the process.

Cautionary Example. As people start to document their processes, you need to keep an eye on them to ensure that they are documenting the processes *as they exist*. It is much too easy, and therefore rather common, for people to start out documenting the current process, and lapse into documenting the process *as they wish it were*. This is nice—there certainly is value in understanding what the ideal process looks like. Unfortunately, to do this is of little or no value, for two reasons. First, you really need to know what people are doing *now*—what process they are really following. Second, if you want to improve your processes and move toward these ideal processes, you need to know where you are starting from. Otherwise your improvement work may go for naught. You will know exactly what I mean if you have ever tried to "fix" a design only to discover that the drawings on which you based your fix were not up-to-date, and the features you changed actually did not look like you believed they did. Your work has been wasted.

Continue with what each process "begins with" and "ends with," the inputs and who will supply them, and the outputs and the customers to whom those outputs go. Do the inputs expected from one supplier match the outputs that supplying process expects to deliver to its customers? Now fill in the details. Use the charting method from Chapter 4 that you and your team prefer. Go only to a level of detail that enables process users to see what steps must be carried out in using the process.

Are there any especially large and complex sub-processes with many users? This will depend upon the nature of your business. If you are in the consumer electronics business, the circuit design process and the software process may be the most involved, and be used by many people. Perhaps these processes should have their own process management teams (PMTs), operating under the leadership of the PMT managing the overall process used to carry out the pilot project. If you form such sub-process PMTs, who should be on them? Who should lead them? If you formed the overall PMT wisely, with a lead person from each process on it, these lead persons are natural candidates to lead the sub-process PMTs.

Each PMT that is organized should meet regularly to track progress, share successes, and surface issues. The achievements of the PMTs can be socialized in a special newsletter for the pilot project, or in an existing company paper.

Put the PMTs to Work

In addition to documenting their processes, there are a number of other things that your PMTs can do. A PMT is in an ideal position to establish process measures, preferably based on internal customer needs. We have already discussed in the preceding chapters several things you might want to measure. Be sure that you are able to collect the data needed by the chosen measures.

A PMT may be the right group to carry out some benchmarking activity.

Each PMT can work to establish requirements for the suppliers to their process, that meet the PMT's needs and are meaningful to them.

You may want to challenge the PMT to create something symbolic of the things they are doing to move to a new way of working. They could create an emblem that could be placed on notebooks or even polo shirts, or perhaps a motto or slogan. The idea is for the people engaged on this pilot project and the related process management work to be bound together, encouraged, and distinguished among their peers. A PMT that we organized early in our work created a variety of things, including specially-printed pads of Post-it notes and the tandem bicycle logo we illustrated in Figure 9.8. The first set of note pads had the word *Process* printed on them. The second set were more elaborate. They were note "cubes" with printing on both sides of each sheet, and on all sides of the cube of padded notes. Figure 12.1 shows the message on the front of each note sheet; on the back we listed the steps of our chosen process management methodology (PQMI).

The logo was developed by the PMT. They had been challenged to achieve interval reduction by cutting the number of prototype cycles from around 3.5 to one, so they adopted as their "rallying cry" *one cycle*. Seeking a symbol to represent this, they were close to choosing a picture of a unicycle. Focused efforts were simultaneously under way to build with our manufacturing counterparts the sense of being a single team; this gave someone the idea of changing the message to *one cycle, one team*. From this emerged the idea of using the tandem bicycle as the logo.

Socialize the Process

As we have said repeatedly, it is important to engage everyone in the organization in your process work and your change to a new way of working. You can begin this by sharing everything that you are doing with the entire organization—all the potential process users and their coaches. You might even get some unexpected gains, since some other folks may have insights into something you may have missed.

We have already mentioned using newsletters, new or existing. Post the processes in high-visibility areas, on bulletin boards or walls. As you gather data, post them on bulletin boards or walls or the doors of the people who have gathered them.

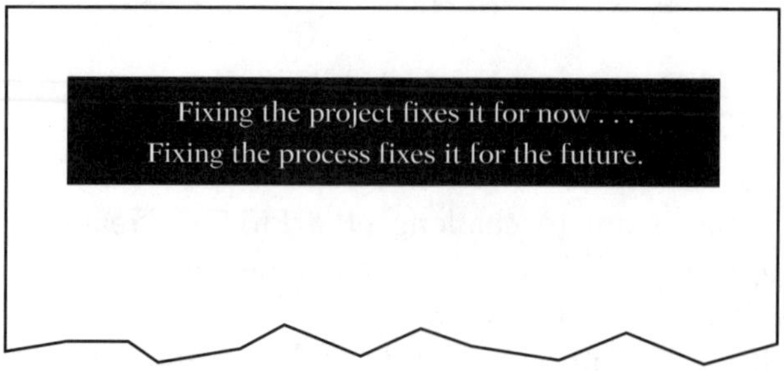

FIGURE 12.1 *Message on the Front of Our "Inspirational" Note Cube Sheets*

To make all of the documented processes available to everyone who may need them or become interested in them, set up a Process Library. At first it may only need to be in thc corner of a lab, or of the office of one of your core team members. You can also use this facility to make available books and articles on process management and improvement. If, as you do early training, you acquire some videos, also put them in the library. Let the library become a gathering place for people who want to know more about your process work.

Example. We have called our library the *Process Resource Center.* It occupies only half of a two-person office, the other half of which is office to one of our process core team associates. The main features of the room that make it our Resource Center are several sets of floor-to-ceiling shelves, a water cooler, a conference table, and a phone line for a video telephone. On the shelves we have put reference books, including some texts and AT&T's "Best Practices" and "Quality Library" books[2], training materials, videos (both motivational and training), and many binders prepared by various PMTs describing their processes. The water cooler and the presence of our team member to provide assistance probably both draw more people to the room than do the reference materials—but once we get them there we have attracted their attention and, we hope, can expose them to more. The conference table is used for a fair number of small process-related meetings, which further enhances people's interest in the Center.

Launch the Project

You need also to get the pilot project underway. The assigned people will start out using the existing processes, using whatever sorts of teams they have chosen to use, and applying their skills to the job. One thing we found to be very helpful before really getting started is for the process coach for the project team to administer and then review with the team a *Pre-Project Readiness Assessment* or *Audit*. This is essentially a final "pre-flight" check by the team that they have the needed re-sources and that the team members are ready to go. Figure 12.3 shows a form for this assessment that we used with some success. Each team member privately completes the assessment. Their scores for each question are averaged by the process coach, who then uses the averaged results as the basis for a "readiness" discussion with the team. There is no "passing" or "failing", and unless some result alerts the team to some particularly critical shortcoming, undergoing the assessment would not delay the

2. Some of these are available to the public. Contact AT&T's Customer Information Center in Indianapolis, Indiana.

Pre-Project Readiness Assessment.		
Score each question on a scale of *1* (Strongly Disagree) through *5* (Strongly Agree)		**Score**
People	Resources are available at project start.	
	I have sufficient time allocated to complete my responsibilities.	
	I feel motivated to make the project successful.	
	The project has a capable, right-sized team responsible and committed to the project.	
	The project team is focused and well organized.	
	Alignment exists among upper management from the various cross-functional areas.	
Process	Team members have process and tools capable of producing expected output, and are skilled in their use.	
	The project team understands the overall development process and their respective areas of responsibility.	
	Key project metrics include measures for project intervals and meeting commitments.	
Project	The project content is appropriately sized for the determined interval.	
	The project will start on time with a complete and stable definition of the project and expected output.	
Platforms	The project does not include "invention" of new hardware, manufacturing technology, processes, or tools.	

FIGURE 12.2 *Pre-Project Readiness Assessment*

start of the project. What it can do is give the team a feeling for their weaknesses, where they will need to focus extra attention to ensure a successful project without panic situations.

The implications of the last question of the readiness assessment about "not including 'invention'" are discussed at the very end of this chapter.

LOOKING AT THE BIG PICTURE

The things we have talked about so far in this chapter have focused very much on the product development community, and on that community's quality issues and their individual functional processes. But if you have read the preceding chapters carefully, you may well be concerned that this is a limited view—that you need to be including your primary customers, who are probably manufacturing engineering, in your quality and process work. You may also feel a need to address the interfaces between functional processes. Only by working with this big picture can you hope to reap the maximum benefits of your focus on quality and process management and using concurrent engineering.

Someone Must Be Accountable for the Overall Process

Let's take the simpler of these two big-picture needs first. Your individual *functional processes* are really sub-processes of your overall product realization process. To achieve maximum interval reduction and concurrency and resource utilization, it is the complete PRP that you need to improve. Remember: sub-process improvement is not enough. We addressed this potential problem in Chapter 4. Uncoordinated low-level improvements can lead to high-level catastrophe: shorter intervals on sub-processes could actually increase the overall interval.

What this probably will require is that you form one more PMT, whose members are responsible for managing the overall PRP. We suggest that the members of this PRP PMT be the leaders of the functional process PMTs. They probably have the best overall view of their functional processes, and of the driving forces and issues that have shaped those processes. They are thus probably the best people to look at how their functional processes play together, and what sort of interfacing issues ought to be addressed.

As an interfacing issue is identified, the wisest way to address it is probably for the PRP PMT to sponsor a Quality Improvement Project or Quality Improvement Team to find the root cause and develop a solution. These QITs or QIPs should include representatives from all the functional processes that have been identified as having stakes in the particular issue being addressed.

Address the Interface with Your Primary Customers

We have already discussed how we changed the relationship between our development organization and our manufacturing engineering colleagues. You may want to do this.

As we explained earlier, before we started this work the relationship between development and manufacturing was quite tense. For mostly historic reasons, there was little positive contact. The two organizations had very different goals and measures. The closest we had gotten together was for the factory to insist that the developers "pass" a DFM Checklist Review before turning a new design over to the manufacturing engineers. We deliberately set about changing the development community's image of manufacturing from an isolated, unknown entity to an integrated partner. We made a conscious effort to change our development and manufacturing processes from two distinct, sometimes conflicting activities, to a single, seamless process with no hand-offs. We changed the roles of our people from each individual working on their own part of the job to concurrent engineering teams focused on the overall process. We began to measure how the manufacturing engineers perceived their development colleagues. This had interesting consequences, described in the following example.

Example. The development organization established a simple survey, in which the manufacturing engineers were asked to rate their development colleagues on several measures. This survey is cited in detail in Figure 7.7.

The early results led us to establish two QITs, to look into how we might improve our *responsiveness* and our management of *commitments*. Each QIT included people from both communities, so that the findings and results would be jointly owned. The outcomes were not obvious. For example, one change that we made to improve our meeting commitments was to get everyone to agree that no person (especially a manager!) could make a commitment for another person. In this way, a person who was committed to some task would *know* that the commitment existed—which was not always the case before. This may seem obvious, but . . .

A second outcome was that after several months, the manufacturing folks decided that they should also ask us how we perceived them, and set up their own version of the survey. It was clear that this activity had led to significant improvements in our working together.

We suggest that you may find building similar relationships between development and manufacturing useful to you, as part of your internal-to-development change to working with a focus on quality and processes. They should bring you even greater gains in interval, cost, and quality.

You may even want to try forging a Relationship Values statement, like the one described in Chapter 7 (Figures 7.10 and 7.11).

It Is Hard to Schedule Invention

Here is another interesting lesson we learned: *do not invent during development.* You may want to consider the possible impact of this as you start up your work to achieve a focus on quality and process. It may not become obvious until you begin looking at the big picture. When the project leader begins to merge the schedules of the individual functional process users and to form a complete project schedule, it may be the first time that anyone notices that the circuit designers are counting on a never-before-tried circuit concept. Will they have it ready on time? Are they sure there are no hidden traps that could delay the delivery? Are the components all available? Are their descriptions correctly entered in the CAD library?

If you are truly committing to your schedule, can you afford these risks? Or is it better not to schedule this invention, but to stick with a known way to achieve the same performance?

When you first mention this to developers, they are likely to behave as ours did and "pooh pooh" the whole concept. After all, isn't invention the real goal of every red-blooded developer?

There is another, better way to manage invention. It is called by many people *Platforming*. The idea is that you carry out your invention—the technological leaps that will keep you ahead of your competition—off the critical path of a committed schedule. Some portion of your development staff, including some of the most innovative and perhaps also people who have just come off the most recent product development project, must be dedicated to this work—on an ongoing basis. They continue to pursue various new ways to meet your customers' needs. When it is certain that some new approach will work, by the platform folks having simulated it, built it up and tested it, and whatever else is felt to be necessary to achieve the needed level of confidence, this new *platform* is made available for use in a scheduled project. In this way, you introduce the new ideas that keep you a technological leader, but with low risk and without jeopardizing committed schedules.

People have found another gain from working this way: shortening the interval for a committed development. If you insist that only proven platforms can be used in committed, scheduled developments, then you can generally complete the development in less time than if you had to build in time to develop and debug the new capability. The total effort needed to introduce some new capability may not be reduced, but once the capability is proven as a platform, the interval from promise to delivery of a product incorporating that capability can be shorter.

Since development engineers will undoubtedly continue to crave the heady experience of inventing, you may want over time to give all the engineers in the organization opportunities to work both on developing new platforms and on designing and developing products. Indeed, a person who has perfected a new platform may want to help incorporate it into a new product.

SUMMARY

If you have kept up with us, your activities implementing your plan to move your organization to the focus on quality and process management should now be well launched. You can look forward to the pilot project being completed, and to metric data beginning to be collected.

In the next chapter we look at activities you will want to undertake to engage more of the organization in the new ways of working, and to begin to see real improvements in the ways you do things.

13 Total Quality Management

Following as it does the chapter on "Getting Started," this chapter could well be titled "Staying the Course." Here we talk about managing your quality programs, and over time applying the quality techniques we have presented.

Any quality program left to run its own course will eventually fade away. Constant nurturing is required. Moreover, your total quality system will most likely consist of a suite of programs to help you reach the Quality Ideals presented in Chapter 3. Some of these programs will do well; others will not. Your ability to adapt continuously to these successes and failures will make the difference in your ability to implement a Total Quality Management (TQM) system. In this chapter we talk about building a lasting environment for your quality system by further

- Improving your Total Quality Management approach (managing the combination of programs you have elected to implement, and building in a continuous improvement cycle)
- Building the overall infrastructure for quality and process management

TOTAL QUALITY MANAGEMENT

Recall in Chapter 3 we presented the following Quality Ideals.

1. Clear Vision and Goals Communicated to All
2. Total Employee Involvement and Empowerment
3. Customers as the Key Drivers for Change

4. Focus on Process: the How of Doing Work
5. Metrics for Continual Performance Assessment

To build a complete quality system within your organization, you need to evaluate the programs you have already started, as well as the ones you intend to implement, against these Quality Ideals. You may find it helpful to develop a matrix to show the mapping between your programs and the ideals (see the example in Table 13.1). The purpose of this table would be to help answer questions such as:

- What gaps do we have?
- What Quality Ideals are only partially addressed by these programs?
- What additional programs need to be considered to round out our quality system?

In those areas where you have incomplete coverage or outages, you may want to consider programs that meet your particular needs.

Next, you need to develop a method to manage your quality system over time. We recommend a Plan-Do-Check-Act cycle for managing your quality system. Start by developing long-term plans for your quality programs. You may have elected to start only a few programs, based on reading Chapters 11 and 12. Now you need a long-term plan for your overall quality system. Using a table similar to the one in Figure 13.1, you can map out your high-level plans for each program as shown in Table 13.2.

As your plans are carried out, you can check on the effectiveness of your quality system by

- Listening to your customers and associates
- Measuring your performance against objectives
- Assessing the success of individual programs

Depending on the type of programs you have in place, this checking step can happen in many ways. For example, to assess the effectiveness of a process management program, we have found it helpful to use a process deployment assessment (see Figure 13.1). This type of measure, taken periodically, will assess your organization's progress in this area. We have elected to use a radar chart as shown in Figure 13.2 to show our progress from one sampling period to the next.

TABLE 13.1 *Mapping Your Quality Programs Against the Strategic Quality Ideals*

(List the Quality Programs you have selected below, and indicate the ideals supported by these programs—see sample list below)	**Clear Vision and Goals Communicated to All**	**Total Employee Involvement and Empowerment**	**Customers as the Key Drivers for Change**	**Focus on Process: the How of Doing work**	**Metrics for Continual Performance Assessment**
Policy Deployment	●		⊙		⊙
Self-Directed Teams for Product Development		⊙			
Concurrent Engineering			⊙	⊙	
Employee Satisfaction Surveys		⊙			⊙
Customer Satisfaction Surveys			⊙		⊙
Process Management Teams for Hardware and Software		⊙	⊙	●	⊙

● Fully achieved
⊙ Partially achieved

TABLE 13.2 *Developing Long-Term Plans for Your Quality System*

(List the Quality Programs you have selected below, and indicate the key objectives at 3 months, 6 months, and so on)	3 Months	6 Months	12 Months	18 Months	24 Months
Policy Deployment	Share program with senior management team	Senior managers attend Policy Deployment training, collect necessary input data	Goals established and fully deployed to organization	Measurable progress achieved in key areas	Repeat process
Self-Directed Teams for Product Development		Identify pilot project; train pilot team	Conduct on-going assessment of self-directed team effectiveness	Deploy on all product development projects	
Concurrent Engineering	Conduct feasibility study per process re-engineering process	Establish re-engineering team and initiate project	New process deployment underway		Measurable improvements visible from new process
Employee Satisfaction Surveys	Develop survey	Collect survey data; use as input to Policy Deployment		Re-survey employees	
Customer Satisfaction Surveys	Develop customer surveys	Collect customer survey data, use as input to Policy Deployment and process management		Re-survey customers	
Process Management Teams	Implement trial with software team; send team to process management training		Measurable improvements for software; assess readiness to extend to hardware	Extend training on process management to all product development organizations	Process management fully deployed to all functional areas, key improvement objectives in place

Process Deployment Assessment

Your Process

1. Do you know where you can find a current copy of the documents describing the functional process you are using?
2. On your current project, are you following the *documented* process?
3. Do you know if a process management team exists for your functional process?

Process Training

4. Have you received any formal training on how to use your functional process?
5. Have you had training in process management?
6. Have you ever applied process management techniques to any aspect of your process?
7. Have you had training in the problem-solving tools? (Things like root cause analysis, force field analysis, Pareto diagram, and data-collection tools like control charts)
8. Have you ever used any of these tools to solve a problem?
9. Have you had training in the "Seven New Tools"? (e.g., Affinity Diagram, Interrelationship Digraph, matrix diagram, tree diagram, prioritization matrix, activity network diagram, and Process Decision Program Chart)
10. Have your ever applied any of these tools?

Process Improvements

11. Do you know how you might go about proposing, investigating, and implementing improvements to your functional process ? (For example: making suggestions to a process management team, forming a quality improvement project)
12. Have you ever participated in a quality improvement project to improve some part of your process?

Projects and Culture

13. Have you ever been directly encouraged by your supervisor to work on process activities?
14. Are you familiar with your project's quality goals and metrics?
15. Have you read the quality plan for your current project?
16. Do you find the quality plan helpful?
17. Are you currently a member of a multi-functional, self-directed work team?
18. Are you willing to bring bad news about your project to your supervisor?
19. If you bring bad news to your supervisor, do you expect a negative response?

FIGURE 13.1 *Process Deployment Assessment*

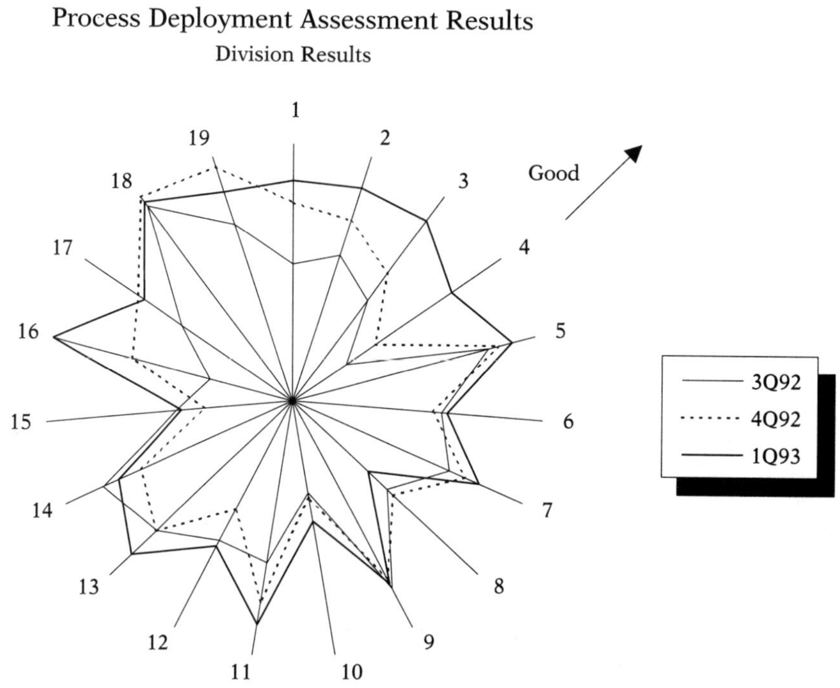

FIGURE 13.2 *Radar Chart of Process Deployment Progress: An Example*

If your process management programs are in place and functioning, you should also be checking the performance data on these processes to see if real improvements are being made.

For policy deployment, you should be conducting annual reviews to check your progress against the objectives that were set in the policy deployment process. You can also survey associates in your organization to evaluate their alignment and support for the key objectives for product development.

For concurrent engineering or other re-engineering efforts, you can check the effectiveness of the process in reducing development intervals, reducing design iterations, improving product quality, or reducing full-stream costs of the product and process.

These various checks or reviews should be complemented by benchmarking other quality systems. You need to keep apprised of what others are doing to implement their quality programs. Where are they having success? How are they doing it? This knowledge is essential if you are to keep pace with the best of your competitors, who are always finding new and innovative ways to improve their products and processes, and better satisfy their associates and customers.

Periodic check-ups and benchmarking can show you where you are doing well, and where more focus is needed. Based on this information, you can re-plan your quality improvement activities. Some programs may need to be re-focused or better staffed, while others may be doing well as they are. You may choose to stop some programs and start others. The re-plan activities should be an open and candid look at what's working and what's not, and appropriate action should be taken.

PDCA Cycle

We have just summarized the Plan-Do-Check-Act cycle for your quality system. Your version of Table 13.2 reflects your long-term *Plan*. The execution of that plan is the *Do* part of the cycle, periodic reviews constitute the *Check*, and the re-plan efforts represent the *Act*. This cycle repeats on a continuous basis as you work your way toward the ideal states represented by the Strategic Quality Ideals. Managing toward these ideals is *Total Quality Management*.

Important to the success of Total Quality Management is the notion of continuous improvement. Continuous improvement means that when the first round is over, the work has just begun. We must constantly evaluate and improve the methods and programs we employ to reach the Quality Ideals, and persistently go after on-going improvements in the product development process.

Our ability to demonstrate continuous improvement in key business measures may determine the longevity of our quality systems. If we cannot demonstrate results, no amount of salesmanship will sustain these quality initiatives. Showing that many activities and teams have been formed and are active may be necessary, but it is not sufficient. We must show continuous improvements in key measures within product development and for the business; namely, improvements in quality, cost, expense, customer satisfaction, and, most importantly, profitability.

BUILDING THE INFRASTRUCTURE

Building an infrastructure that supports Total Quality Management is crucial for the long-term success of your quality system. The focus areas for building infrastructure are:

- Training all associates, first level through top management
- Reward and incentive programs to encourage role models and recognize success
- Associate development

Note that in Table 13.2 we showed several planning steps to do training. Training can be a powerful influence on people to adapt new ways of managing and improving their work. Training should cover specific quality programs as well as the overall quality system. Topics such as effective use of teams, process management, policy deployment, and problem solving are sample topics for a training curriculum that goes beyond the normal skill training in each functional area.

The reward system also needs to be structured in such a way that you continuously drive behavior in the direction of the Quality Ideals. That means rewarding process improvements, team problem solving, metric definition, data collection, improvements in customer satisfaction, and so on. Most importantly you need to reward role-model behavior in these areas. Rewarding those who set good examples for others helps propagate your system throughout the organization.

Over time you will find that you can change the value system of the organization by changing the reward system, and as the values of the organization change, the behavior of individuals changes with it. You will begin to hear people ask:

How do you think other companies do this process?
What data from our customers support this change?
How do we perform against competitor benchmarks?

As you observe these changes, you will also see individuals who, more quickly than others, develop an appreciation of the Quality Ideals and the associated practices and tools. You should attempt to develop these associates into sponsors of the change. Provide them with personal attention and special training. They will magnify your efforts to deploy the quality system.

By continuing to transform the overall culture to support your quality system, and by executing the Plan-Do-Check-Act cycle on your quality system activities, you will eventually bring about a successful change in your organization to a Total Quality Management approach—one that embraces the Strategic Quality Ideals and works continuously to achieve them.

Part IV

LEARNING FROM YOUR EXPERIENCE

14 Celebrating Your Results

Strategic Quality Ideals

✓ Clear Vision and Goals Communicated to All

✓ Total Employee Involvement and Empowerment

✓ Metrics for Continual Performance Assessment

Organizational Enablers

✓ High Morale and Satisfaction of Employees

✓ Reward Structure Linked to Contribution

INTRODUCTION

We have already alluded to the special importance of *rewards* and *recognition* to keep people working on these new sorts of jobs, toward these new types of vision, mission, and goals. In Chapter 8 we discussed how teams need to be rewarded and recognized in new ways, and in Chapter 11 we talked briefly about the value of rewards beyond basic compensation.

This topic is sufficiently important that merits its own chapter. "When an organization consistently and publicly recognizes employees for improvements and quality-related behaviors and achievements, the rest of the organization notes the recognized performance, internalizes it, accepts it as the norm and adjusts behaviors accordingly. More change and improvement ensue" (McArdle 1993, 34).

Engineers are notably poor at *celebrating* their achievements. Many engineers display the attitude that their job is to innovate and achieve, and that there does not need to be any special public acknowledgment when they do their job well. They often appear to derive their greatest satisfaction from solving a knotty problem, and are visibly embarrassed when given unexpected praise. Not that they don't enjoy a celebration when somebody holds one—but it takes an extra effort to make sure that someone thinks to have the celebration. This was brought out to one author very clearly (and a bit painfully). After years of coaching engineers who are relatively blasé about their company Service Anniversaries, he was unprepared for the importance a Fifteen-Year Company Service Anniversary had for an associate with a sales background who had recently joined his group. It was only after she expressed her disappointment that her Service Anniversary day had come and gone without any special words to her from her coach that he realized his goof. The company anniversary gift had come, but it had been delivered right to the associate because the coach's office was in a different location. It was only after extended discussion that he began to understand how important it is to sales people to have their achievements publicly recognized by activities like "Million Dollar Clubs" and "Achievers' Clubs."

The lesson was reinforced by another event that occurred about the same time. In an attempt to integrate the development community better into the business unit, top management proposed to create for the development community a variation on the "Achievers' Club" that appears to be a really strong motivator in the sales community. The engineers overwhelmingly rejected the concept of a few being selected to receive a free trip to a Caribbean Island in celebration of their outstanding technical achievements. They threw away the gimmicks (letters from top managers and a cassette tape) mailed out to launch the program. They voted instead to have much less elaborate local dinners for entire teams who had met their development goals in an exemplary way. Think for a minute: where was your own professional society's last conference—very likely in the home city of the professor in charge of planning the conference. Members of other professions regularly meet in Hawaii or other exotic places; engineers tend to be apologetic if the conference city seems too much of a prime vacation spot.

Discussion with the associate with the sales background brought out the view that this all makes sense when it is related to *psychological type*. Engineers tend to be *Introverts* (type "I"), while marketing people tend to be *Extroverts* (type "E"). Marketing folks, the associate pointed out, are directly motivated by public rewards that demonstrate top management's interest in *them*, and that can be hung on the wall for all to see. People consciously work harder in order to achieve these recognitions. She recalled that when she had received the promotional material for the new award that the engineers had considered to be "gimmicks," she was excited by the opportunity, and told her coach at the time that she was determined to earn one of the awards. This is directly supported in a discussion of psychological type as determined by the Myers-Briggs Type Indicator: "Extroverts, for example, typically need a great many more overt 'strokes' than Introverts. Introverts, in contrast, tend to be suspicious of those same overt 'strokes.' True, both need affirmation, but too much affirmation makes the Introvert wonder why so much is needed, while to the Extrovert 'too much affirmation' is a contradiction in terms" (Kroeger and Thuesen 1988, 35).

Perhaps the most important point is that *both need affirmation*. The engineers *need* to have their achievements celebrated; they may need prodding and support to remember to celebrate one another's achievements.

As we work to move all of our associates to the sort of new environment we have been discussing—to a new culture focused on quality and on *how* we do things and *who* does them, rather than so strongly and narrowly on *what* we do—it becomes more important than ever for everyone to be recognized. The nature of each person's job is different, the "product" of the job is different (you cannot hold and admire a new process in the same way you can a new product), and so the way the achievement is recognized and rewarded needs to be different.

USING REWARDS TO DRIVE DESIRED BEHAVIOR

This may sound mercenary, but it is a very important, practical consideration. People tend to behave in order to maximize their rewards. A particular person's sought-for reward may be a large monetary prize in one instance, the satisfaction of a job well done in the next, and the warm feeling of having brought pleasure to a loved one in the next. In each of these instances, the reward makes a strong contribution to the motivation that drives the individual to complete some challenging activity. Over the years we have seen that when a salesperson has a mix of products to sell, the details of the compensation plan (i.e., how much commission they receive for selling each different product in the mix) will very strongly determine which of the products the salesperson will work hardest to sell.

As you work to move your organization to a new focus on Quality, you want people to behave in new ways, which often force them beyond their familiar comfort zones. You want them to challenge their habits and their established ways of doing their jobs. You want them to work in new environments. It stands to reason that you need to celebrate their accomplishments and reward them in some new ways, and perhaps more often than they are used to, in order to reinforce the desired behavior and keep them motivated to continue to move toward the new vision.

The types of rewards and recognitions you may want to consider can be classified in these categories:

- Special compensation
- Awards
- Celebrations
- Mementos
- Other activities

In the following sections we suggest some things you might do in each of these categories.

SPECIAL COMPENSATION

You may want to consider some different ways to compensate people as your new culture develops, beyond the conventional R&D combination of salaries and raises, and bonuses, or "lump-sum awards," given on the basis of individual performance. The new environment of people working in teams, especially self-directed teams, may be better supported by having at least some part of compensation based upon individuals' performance as members of teams. We have seen and heard of several ideas people have tried; those set out here seem to have worked.

Bonus for Achieving a Goal

Some part of the organization's funds for annual bonuses is set aside, to be distributed to teams when they complete projects or achieve significant intermediate milestones. The money is given to all team members, either as equal shares or according to some formula that perhaps gives larger shares to professionals than to technicians. The same people might have received this bonus money anyway, but now the presentation is linked to their performance on the team. Each individual's share may be different than it otherwise might have been, since it is based upon team performance rather than upon individual performance.

Annual Raise and/or Bonus Based Upon Team Participation

Rather than basing people's annual raises or bonuses upon their individual performance, the members of a team are given their raises and bonuses based upon their participation on the team. There may be no special tie between raise or bonus and the team achieving any particular milestone, but the message is made clear that the performance of the team has an impact on the amount of money available for raises and/or bonuses, and being a member of the team has determined each individual's share. One way to divide up the money is to give "equal shares"—where the shares may be *equal amounts* or an *equal percentage* of each individual's base.

Team-Determined Shares

In this arrangement, the team members decide whether the shares will be equal, or if some individuals will be singled out for either outstanding performance or poor performance, and their shares adjusted accordingly.

Peer Awards

An amount of bonus money can be set aside, to be given out at some preselected time during the year on the basis of peer nominations. All noncoach (nonmanager) associates are invited to nominate colleagues for special extra bonuses, based upon some special contribution to the team or organization. A designated review committee gathers the nominations and selects the award recipients based upon frequency of nomination and/or the nature of the contributions. A second set of nominations may be solicited, for distribution of additional funds set aside for the "outstanding first-level coach."

All such plans must be carefully administered to ensure that the entire organization feels that the raises and bonuses given in these ways have been determined fairly and without prejudice. Team awards may be difficult in a mixed environment, where some people receive their raises based upon individual performance and others for their performance on teams. If the goal is to convince people of the value to the company of taking the risk of working on a team, the team-based raises and bonuses may want to be generally higher than those to non-team participants.

AWARDS

A very clear way to signal that certain individuals or teams have made outstanding contributions to the new way of working is to publicly present them some sort of *award*. Such awards may have any of several forms, including

- *Monetary*, ranging from a special bonus to a Gift Check or an opportunity to voucher an evening out with one's significant other
- An *artifact with intrinsic value*, such as a cut-glass paperweight or candy dish, or an expensive pen set
- An *artifact with no intrinsic value*, such as a plaque or a framed certificate
- A *special privilege*, such as the use of a special reserved parking space for a month

The size or value of the award will vary according to the organizational level at which it is presented, and the difficulty of the requirements for the presentation. Thus it would be expected that an award to be presented by the R&D organization's Executive Quality Council would be given only for a truly outstanding contribution and the artifact would be clearly distinguished, while an award that could be presented by a process management team leader would normally be given for a contribution on a much smaller scale and would have little intrinsic value.

Major Awards

The following paragraphs describe some specific examples of major awards the authors have observed being used.

Division Quality Award

For an R&D organization of about three hundred people, a team award was created, consisting of a personalized cut-glass crystal paperweight (Figure 14.1) and a framed photo of the team, to be presented to each member of the team. Nominations were accepted at any time from anyone in the organization, for any team that was felt to have made a visible contribution to quality or to process improvement. The organization's Quality Council evaluated the nominations, and twice a year presented an award to a selected team. Two examples of the teams recognized were

- A pair of associates who published a newsletter for members of a major R&D-plus-manufacturing process management team
- The team of coaches who, over time, had significantly simplified and improved the organization's annual performance review process

Company Team Award

Once each year, the organization published guidelines and solicited nominations for an organization-wide team award. Several awards were made, each comprising a single plaque naming the team and all of the members, plus a personal letter to each team member from their corporate vice president. The plaque was designed to be displayed in some central place in the work location.

Corporate Quality Award

AT&T Bell Laboratories several years ago instituted the *President's Quality Award*, which is presented annually to one or more teams that have made significant contributions to Quality in the AT&T R&D community. This award consists of a very significant trophy that is normally kept at corporate headquarters, but that rotates for a time among the award recipients' work locations. In addition, each member of each winning team receives a framed certificate. The award is presented each year at the Bell Labs Quality Day.

FIGURE 14.1 *Quality Award*

Lower-Level Organizational Awards

Smaller organizations can also create their own awards, even when there is also a major company award program. In the smaller organization, the nomination and selection process will normally not be as complex. Often the deserving team or individuals may be identified by their own coaches, and the award conferred by the top manager of the organization. Some examples of quality awards in this category follow:

Team Dinner

Management might take an entire team, perhaps with their significant others, to dinner at a nearby fancy restaurant.

Team Pictures Prominently Displayed

During a benchmarking trip to another company, the authors were shown a wall near the company cafeteria on which were posted group photos of teams that had been recognized for their outstanding contributions. Some of the teams had been recognized for *project* achievements, and others for *quality* achievements.

Public Recognition

Another way to recognize outstanding contributions to quality and process improvement is to publish their stories regularly in the company magazine or newspaper, or an organization newsletter. This could even be done in a regular column, so that associates come to look to see who has been recognized for what. These stories could be in conjunction with other awards, or stand alone.

Local Awards

At the lowest (but to some people, the most important) level are team or individual awards presented by team leaders or coaches when the team or some team member has done something of special note. These are "low-energy" awards that do not require extensive deliberation or approval. Some examples follow:

"Dinner-for-Two" Awards

Let first-level coaches know that they can give out a reasonable number of awards consisting of approval for each individual being recognized to go to dinner with a significant other, and voucher the cost of the dinner (up to a specified amount, perhaps $125). An award like this is normally conveyed via a letter from the coach or team leader. In case the person does not particularly want to go to dinner, the award might also be used for theater tickets, or perhaps put toward the cost of something like a new CD player.

Peer Recognition

One coach with whom we are acquainted established a very simple peer recognition program for his organization. He allocated $100 to each associate in the organization, to be used at any time during the year to recognize any other associate(s) in the organization. The recognition, called the *"I Really Appreciate" Award*, can be for teamwork on a project, on some activity for the good and welfare of the organization, on a community service project, etc. It can be in the form of a gift or of cash; a copyable certificate is available if the person making the recognition wants to use it. The mechanism for recognition is absolutely simple: just do it. Afterward, the person making the recognition has only to fill out a simple form and give it to the organization's secretary to be reimbursed from their $100 allocation. The only rule is the $100 annual limit.

How successful has it been? At the end of the first year, the coach called it one of the best-spent $5000 of the year. Almost everyone used up their allocation. One person gave a $100 gift to one colleague; another person purchased 100 dollar-boxes of candy, and gave them out one at a time. The organization perceived that there were no losers, only winners (although someone who was never recognized probably got a message . . .), and they voted overwhelmingly to continue the program. The certificate was made available on personal computers, and "they are all over people's walls." Tremendous value has been obtained from a very simple recognition program.

Simple Collectibles

In Figure 8.3, we illustrated a Team Award magnet devised for one of our major process management teams. (It is part of our company culture to use magnets to fasten various things—usually papers—to our metal wall panels and modular furniture.) Bearing an emblem previously devised by the team as their logo, the magnets were kept by the PMT leader, to be awarded at the suggestion of some coach or subteam leader. Since the magnets were useful, multiple awards to the same individual over a long period of time would not diminish in importance by seeming silly or becoming lost in desk-top clutter.

Simple as the magnets are, a number of recipients were really appreciative of being recognized by their team leader for a special Quality contribution.

The idea for these magnets was derived from an experience of one of the authors when visiting a vendor's factory. He observed over some workplaces a "chain" of brass stars, engraved with accomplishments and dates. The story was that when an individual worker significantly exceeded a weekly production goal, or when a team completed some especially difficult order on time, stars were awarded. It became part of the culture to hang them above the recipient's bench or desk, in a chain. While these were awarded for product achievements, it seemed reasonable to copy the idea for quality achievements.

CELEBRATIONS

Another way to recognize teams, especially larger teams and teams who will be continuing to work together over an extended period of time, is to celebrate together.

When a team achieves a significant goal, such as completing a project using a newly improved process with a reduced number of design iterations, it might be appropriate to take the team out for pizza. This both allows members to "feel good" together and demonstrates management's appreciation of their achievement.

This is also a good way to recognize an achievement by the behind-the-scenes quality support team—your own core team of quality professionals. Let's say they have been coaching the organization to prepare for an ISO 9001 certification audit. The organization achieves certification, and everyone celebrates. It might be good for the morale of the support team also to treat them to dinner as a team, where they can reflect on their success in coaching the larger organization to meet this important goal.

If it is not possible to identify a single particular team achievement worth celebrating, a team that has, over the course of a year, established a record of continuous improvement might appropriately share a holiday or year-end luncheon. The authors did this one year for a PMT with several members at each of two widely separated locations. We held a pair of nearly identical luncheons, two days apart. Two of us emceed the lunch party in New Jersey, and then flew to Denver to do the same there two days later. At each luncheon we ran a fun ice-breaker game, presented homemade award certificates to significant contributors, and thanked *everyone* for their contributions.

Another quite different way to celebrate a team's achievement might be to arrange a visit to the team by a company leader they don't usually get to meet—perhaps your vice president, or even the company president. The team members are likely to feel really valued if this significant person in the company takes the time personally to thank and congratulate them for their accomplishment.

Don't forget to celebrate individual team members. At the Milliken Company, an early Baldrige Award winner, "TV screens at each site daily scroll the names of all associates having birthdays that day" (Myers 1990, 17).

MEMENTOS

Not all recognitions need to be for specific achievements. It is sometimes appropriate to recognize participation on a team by presenting everyone with a memento. They might be presented when the team's job is done, or at some intermediate point when the team is at a psychological low point and could use something to build up spirit for an all-important final push. Typical mementos, which are usually customized with some team-developed logo or emblem, are

- T-shirts or polo shirts (Note that the people modeling the "role" hats in Figure 8.4 are wearing polo shirts with the "one cycle, one team" logo.)

- Pins (The staff in AT&T's factories have a culture of wearing such "participation" pins on their corporate identification badges. Recently these have been given out for plant-wide quality initiatives.)
- Coffee mugs
- Paper weights
- Coasters
- Team photos for everyone

Mementos such as these have the added value of publicizing the team or activity to all who see them. This can help to build the interest of other members of the community in your quality and process improvement efforts.

OTHER ACTIVITIES

Milliken uses another type of activity, called a *Sharing Rally*, both to recognize people who have made special improvement achievements and to spread the winning ideas around the company. Innovative associates from all sites are brought to the rallies, which are held at company headquarters each quarter, to present their improvement ideas. Top corporate management participates to show the importance of these special activities (Myers 1990).

SPECIAL REWARDS FOR THE QUALITY/PROCESS TEAM

Don't forget to recognize the folks who have worked hard to make your quality system work. It is important that you celebrate your own successes and show your appreciation. This may be especially important if the company tradition has given (and still gives) the most credit to those who directly contribute to product development. Any of the sorts of special recognitions suggested for development teams also apply to this special team.

If you can arrange it (or if it should happen spontaneously), recognition by the project teams they are serving may be the most meaningful expression of appreciation to these folks.

SUMMARY

This has obviously been only a partial introduction to the concept of new kinds of rewards and recognition. While we have made a number of suggestions from our own experiences, the sky is (more or less) the limit for creating your own—within the context of your organization's culture and the available funds. The important thing is clearly and repeatedly to reward and recognize people for demonstrating that they are adopting the new focus on quality, espousing the new culture, and successfully achieving results in the context of your vision, mission, and goals. As we said earlier, err on the side of too much recognition rather than on the side of too little.

Epilogue

In the last fourteen chapters we have discussed methods to improve product development productivity. These methods are based on some common sense ideals including:

1. Clear Vision and Goals Communicated to All
2. Total Employee Involvement and Empowerment
3. Customers as the Key Drivers for Change
4. Focus on Process: the How of Doing Work
5. Metrics for Continual Performance Assessment

Even when specific quality programs come and go, and buzz words go out of style, these principles or ideals will see little change. We will always look for clear visions and goals to guide our work, and rely on employees to help improve effectiveness and efficiency. We must continuously adapt to the changing needs of our customers and tactics of our competitors, and we will always need to measure our performance.

Depending on the business climate, process management as discussed in Chapter 4 may be all you need to employ right now. Process management and metrics are the most fundamental elements presented in this text, and continuous improvement is built into these concepts.

Sometimes business needs or severe performance problems will force you to re-engineer a process from scratch to achieve breakthrough improvements, as described in Chapter 5. This approach is becoming more and more popular for product development processes because of the fast turnaround and the "results" focus of the work.

Continuous improvement is needed for problem solving as well, as discussed in Chapter 6. More and more employees will require basic problem-solving skills, especially with the growing popularity of teams in the workplace. Continuous improvement in this area means on-going training for new employees and refreshers for everyone to keep these skills honed and fresh on the latest tools and techniques.

Concurrent engineering's day has come. If you have not employed this very powerful concept as discussed in Chapter 7, then you cannot start soon enough. The savings associated with CE are some of the most well-documented benefits of applying process improvement between R&D and manufacturing.

More and more the use of teams invades the work place. Fast decision making and effective communications throughout the organization are critical success factors. Cross-functional teams, self-managed teams, quality improvement teams: the list goes on and on as described in Chapter 8. Continuing education on how and when to employ these different teams can make the difference between successful and unsuccessful quality and productivity improvement programs.

Knowing what to work first—establishing priorities for improvement work and setting objectives, and deploying these goals throughout the organization in such a way that each individual decision supports the objectives—is essential if your product development organization is to improve year after year, and gain breakthrough improvements in key areas. When these objectives are based on customer input and competitor information, you have the basis for Policy Deployment as discussed in Chapter 9.

These concepts and tools may seem like plain common sense to you. Your challenge may be, as it was for us, to pull it all together and implement these concepts throughout the organization in a sensible way, as discussed in Chapters 10–13. Whether you are faced with difficulties getting funding, a traditional culture that is reluctant to change, or a few strong nay-sayers who are standing in the way, be assured that others have worked through similar problems and have deployed an effective quality system.

At this point we have come full circle, and will close. We have set the stage for productivity improvement by discussing concepts, tools, and application of these tools for product development. It is now your turn to make these ideas a reality for your organization. We hope your journey produces results for your organization, and a wealth of learning for you.

Appendix: The Quality Improvement Story

Chapter 6 develops the tools we have found useful in applying process in a development environment. One of the tools that we referred to in that chapter is the Quality Improvement Story (QIStory or QIS), originally developed by Florida Power and Light, Inc. We have repeated the process here in tabular form for reference. The QIStory, sometimes referred to as the *Quality Improvement Cycle*, or *QIC*, is based on training and processes licensed from Qualtec Quality Services, Inc., a Florida-based international consulting and training organization.

TABLE A *Quality Improvement Story**

Step	Objective	Key Activities	Tools/Techniques	Outputs
1. Select improvement area	Identify a problem area and the reason for working on it	1. List possible problem areas 2. Collect data 3. Select problem area for improvement 4. Schedule the QIS activities	Brainstorming Multivoting Interviewing Checksheet Interviewing Team Project Planning Workshop Problem Selection Matrix Pareto Diagram Graph	1. Problem area and rationale 2. Team Project Planning Worksheet for QIS schedule
2. Identify outputs and customers	Identify outputs and customers	1. Identify outputs 2. Identify customers	Customer/supplier Model	1. List of outputs 2. List of customers
3. Determine customer expectations	Understand customer requirements and degree of satisfaction	1. Identify characteristics, customer requirements and priorities 2. Establish measures and collect data 3. Assess conformance to requirements and describe the gap	Customer/Supplier Model Customer Needs Analysis Survey Customer Needs Analysis Checksheet Graph Control Chart	1. Description of the gap 2. Customer requirements and satisfaction level
4. Describe current process	Understand how things actually work and the performance necessary to meet customer requirements	1. Identify inputs and suppliers 2. Flowchart the process 3. Collect data 4. Assess process performance and establish targets	Customer/Supplier Model Block Diagram Flowchart Checksheet Graph Control Chart	1. List of inputs and suppliers 2. Flowchart of process in question 3. Targets for overall process performance
5. Focus on improvement opportunities	Select opportunities to pursue	1. Identify contributing factors 2. Select opportunities to pursue 3. Write the problem statement	Pareto Diagram Multivoting Problem Selection Matrix Problem Statement Matrix	1. Prioritized list of potential problem areas 2. Problem Statement

TABLE A *Quality Improvement Story (continued)*

Step	Objective	Key Activities	Tools/Techniques	Outputs
6. Determine root cause (diagnostic journey)	Identify and verify root causes of the problem	1. Perform cause and effect analysis on the problem to the level of root cause	Brainstorming Fishbone Diagram	1. List of root causes to be addressed
		2. Identify actionable root causes	None required	
		3. Support potential root causes with data for improvement	Checksheet Pareto Diagram	
		4. Select root causes with greatest probable impact	Pareto Diagram Histogram Graph	
7. Trial and implement solutions (remedial journey)	Develop, test, evaluate and implement solutions	1. Develop and select potential solutions	Brainstorming Multivoting Force Field Analysis Solutions Selection Matrix Cost Estimation	1. Implemented solutions and data on impact
		2. Develop an action plan	Action Plan	
		3. Test, evaluate and revise solutions	Checksheet Graph Control Chart	
		4. Implement and monitor solutions	Graph Control Chart	
8. Hold the gains	Sustain improvement, consider replication, evaluate QIS implementation and plan future actions	1. Recommend required education and revised methods and procedures suppliers	None required	1. Control at new performance level 2. Replication recommendation 3. Celebration 4. Plans for next project
		2. Establish tracking system to monitor results	Graph Control Chart	
		3. Recommend areas for replication	None required	
		4. Plan further actions, if necessary	Action Plan	
		5. Evaluate QIS implementation	QIS Review Form Team Project Planning Worksheet	

Glossary

Alignment. The concept that the goals of an organization match those of higher and lower organizations in the same company (vertical alignment), and/or match those of the organizations at the same level in the company, with whom they interact (horizontal alignment).

Associate. Employee. The term *associate* is a applied to *all* of a company's employees, from the CEO to the person carrying out the most menial tasks, to convey the concept that regardless of level, salary, etc., all share in the responsibility of making the company successful, and all share in the rewards of that success.

Baldrige Award. See **Malcolm Baldrige National Quality Award**.

Behavior. In the psychological sense, how a person acts and reacts to their environment.

Benchmarking. Comparing how your company does some job, especially how *well* the job is being done, with how other companies do it. Benchmarking is most useful when the comparison is with other companies generally acknowledged as doing the particular job well.

Best-in-Class. Of a number of similar processes, the one demonstrated to yield the best results.

Best Practice. As applied to processes, typically refers to that process, among a set of processes, that is believed to perform the desired operation the best, namely, with the lowest cost, highest quality, shortest interval, and most satisfied customer.

Brainstorming. A methodology for capitalizing on the dynamics of group interaction to bring forth as many potential solutions as possible to some stated problem. The formalization of the methodology is generally ascribed to Alex Osborn (see Osborn in the Reference section).

Breakthrough Improvement is typically contrasted with continuous improvement, and is a sudden or abrupt improvement in performance levels. Process re-engineering typically produces breakthrough improvements.

Business Case. A description of the impact of a proposed development project on the company's business. Includes expected time and cost to develop, anticipated risks, estimated market data (including expected price, plus total expected sales at that price and perhaps other possible prices), and a summary impact (i.e., expected net revenue). Usually the business case is first prepared during the *Concept Phase* (see below), and refined as the development progresses.

Cascade. A methodology for making everyone in a company aware of some goal, policy, etc., by passing (cascading) the information down through the levels of the organization. The communication at each level is derived from the one above.

Change Agent. A person or team specifically charged with implementing change in an organization.

Coach. One who instructs, trains, or guides an individual or a team. In the modern business environment, the role of the "boss" changes from that of a *manager,* who dictates how subordinates should do their jobs, to that of a *coach,* who guides associates to bring their best skills to bear on the job.

Coach, Process. See **Process Coach**.

Comfort Zone. The concept that, based upon experience, education, personality, etc., an individual has developed a sense for what they feel comfortable doing or attempting to do, and tends to do only those comfortable things. Changing a person's behavior usually involves motivating them to do some new things that lie outside of their comfort zone. Fear, uncertainty, lack of skills or knowledge, apathy, and many other factors can be barriers to changing behavior.

Compensation. The total package of pay, including salary, bonuses, benefits, etc., that an employee receives in return for services rendered to their employer.

Concept Phase. That portion of a development project in which the product concept is documented and evaluated. It may include evaluating the market, ascertaining technical and manufacturing feasibility, and estimating financial impact, cost, and time-to-market.

Concurrent Engineering is an approach to product design wherein all aspects of a product's total life cycle are considered during the design of the product, by a cross-functional design team that works all aspects of product development—hardware, software, test planning, manufacturing process development, etc.—at the same time rather than serially.

This concept began with Design for Manufacturability (DFM) and simultaneous or concurrent design of the product itself and the manufacturing processes that would be used to make it. The scope was expanded by the Department of Defense (Institute for Defense Analysis 1988) to include not only DFM but also other life cycle considerations—indeed, it became "DFX" (Gatenby 1988), where *X* stands for all the other aspects of the product's life cycle (e.g., Design for Serviceability, Safety, the Environment, Disposability). It is not merely doing both the design of a new product and the design of the associated manufacturing processes at the same time. This early, simplistic definition has generally been superseded by the broader definition presented above.

Consultant. An expert, usually outside of the company, who contracts for specialized services not available within the company. Examples include management consultants, quality consultants, change management consultants, etc.

Continuous Improvement. The concept developed by proponents of quality that refers to gradual, continuous change for the better—often contrasted with *Breakthrough Improvement*, which results in sudden, major improvement in performance (see above). Continuous improvement ideally proceeds at a faster rate than the most challenging competitor is improving.

Core Team. In a multi-layered team structure, the core team are those people with the primary responsibility for completing a job. For a product development project, the core team might include one representative each from marketing, development, manufacturing, and distribution, and be charged with the responsibility of making the proposed product available for sale. Each of these people might in turn lead or create other teams to carry out their portion of the work.

Courseware. A set of materials that accompany a training course, to be used by the trainer and/or students.

Culture. Refers to the overall set of values, rules, accepted behaviors, and general rituals within an organization. Often process changes or new quality programs involve a change to an organization's culture, and as a result take a great deal of time and resources to implement.

Customer Satisfaction. A measure of the degree to which a company's customers are satisfied with their products and services. Sometimes measured in terms of a customer's willingness to buy another product from the same vendor.

Cycle. Executing some sequence of work activities a single time. Drafting, fabricating, assembling, and testing a circuit pack would be a single cycle in the design and development of that circuit pack.

Cycle Time. The time required to execute a single cycle.

Deming Prize. An award created by the Japanese in honor of the late W. Edwards Deming for his contribution to the rebuilding of Japan after World War II. The award is given primarily to Japanese companies that demonstrate excellence in their overall quality system. Florida Power and Light was the first U.S. company to win the Deming Prize.

Design Cycle. See **Cycle**, above.

Development is the process of taking a defined product and transforming that product definition into a manufacturable product.

Development Interval. The length of time taken to carry out a product development. Depending upon the company's particular development process, the development interval is usually specified between two milestones such as (a) having an agreed-to set of requirements and (b) approving the product for mass production.

Doers. The people who are charged to carry out ("do") some specific job or set of jobs.

Downstream refers to processes or operations that occur after or subsequent to the process currently being executed.

Drafting. The job of creating a detailed description of some product so that it can be manufactured. Traditionally done by manually drawing detailed pictures using standard symbols and formats, drafting is increasingly being done using computer-based applications. The term *drafting* is also often applied to the organization responsible for doing drafting.

Driver. The "force" (situation, rule, event, etc.) that causes some other activity to be carried out.

Employee Satisfaction. A measure of how employees feel about their job, their work environment, their leaders, their compensation, their opportunities for growth, the company in general, etc. Employee satisfaction is often measured in terms of employee responses to a set of questions about their attitudes toward various aspects of the company and their perceived place in it.

Empowered, Empowerment. Refers to an individual or team believing that they have the power, derived from some higher authority, to make decisions about how they carry out their work. The term is frequently mis-used in the sense of a "boss" *empowering* a team of workers—as though the boss could magically convey the power to the workers. The truth is that while the boss must create the environment for empowerment, the people must decide for themselves that they are empowered.

Environment. The complete set of circumstances and conditions in which an individual or team functions.

Evangelist. An individual whose job is to spread a concept or idea among some community.

External. Outside of the company.

FCC. Federal Communications Commission. The federal agency charged with, among other things, how the electromagnetic (e.g., radio) frequency spectrum is used (and not mis-used) within the United States.

FCC Part 15. A set of rules from the FCC governing what sorts (frequency and power) of electromagnetic waves may be emitted by products designed for use in homes and businesses. The purpose is to ensure that stray emissions from one product do not affect the performance of another—so that, for example, your personal computer does not emit radio-type waves that make it impossible for someone in the next room to watch television.

Front-End Process. The processes before the detailed definition and development of a product or offer. These processes usually deal with strategic directions, market intelligence, customer data, and competitive information. The output of a front-end process is typically a product concept and business plan, and potentially technology platforms for the new product.

Full Stream. When applied to processes, *full-stream* refers to the entire set of processes needed to conceive, develop, produce, sell, service, and maintain a product or offer.

Functional Manager. A supervisor or coach responsible for a group of employees with similar functional skills—for example, software engineers or circuit design engineers.

Functional Process. A process typically used by people with similar functional skills. An example would be a process for carrying out the electrical design of a new circuit pack.

Goal. A desired endpoint for a work effort, usually specific and measurable; also called *objective* or *target*.

Hoshin Planning. A process adopted from Japanese management philosophy, used to establish goals and objectives for an organization and to guide the subsequent planning and execution of activities in a Plan-Do-Check-Act cycle, in accordance with those objectives. Also called *Policy Deployment* or *Management by Policy.*

Internal. Other organizations within the same company.

Internal Customer. A person in the same organization or another organization within the company to whom some output must be delivered. Example: hardware developers must provide models to software developers, so that new software can be tested in a realistic environment.

Interval. The time it takes for a job or process to be completed.

ISO 9000. A set of international standards used to rate the *Quality System* (see below) of a company, and thus the company's ability to execute the conditions of a development and/or manufacturing contract in a quality manner.

Life Cycle. The entire life of a product, from the time it is first proposed until units that no longer function are returned, disassembled, and their constituent materials returned to the resource stream. Often used to refer to the period of normal or standard production, after development, and up to the time when manufacture of the product is discontinued.

Malcolm Baldrige National Quality Award. An award established by the U.S. Congress, given to companies who best exemplify Quality performance, as judged against a comprehensive published set of targets. The award is named for a late Secretary of Commerce who was killed in an accident.

Management. Individuals and organizations in a company responsible for establishing policies and plans for operation of the company.

Metric. A measurement, the variation in which over time can be used to assess the performance of an organization or process.

Mission. The purpose of an organization—its reason for being. Also refers to an organization's primary value-adding functions.

Multi-Function Team. A team whose members represent a number of functional skills. An example would be a product development team with members from marketing, hardware development, software development, manufacturing engineering, factory process engineering, and material management.

Nay-sayers. Those people who do not support the change that is being implemented, and who typically verbalize their doubts.

Norming. A stage in team development where the team members, having worked out how they can best work together, select the methods and establish the rules by which they will do their job.

OOPS. A mistake or error, especially one that should have been easily seen and prevented or avoided.

Paradigm. A pattern, example, model, or archetype; "the way things are" or are done.

Paradigm Shift. A change in the way things are seen or done, often involving trauma for those people with the greatest stakes in the status quo. A paradigm shift forces many people to act outside of their *comfort zones* (see above).

Pareto Principle. Juran has described the principle of separating the "vital few" from the "trivial many" as a means of determining priorities. He provides (Juran 1964, 44) an interesting account of the origin of the Pareto principle and its application to management. Juran states the possibility that in his re-invention of the "principle" about forty years ago, he may have attributed it to the wrong person (Juran 1960). (See also the section on Pareto charts in Chapter 6.)

Performance Review. Formal process, usually annual, for assessing the performance of employees, often for purposes of determining and adjusting compensation.

PMT. See **Process Management Team.**

Policy Deployment. See **Hoshin Planning**.

PQMI. "Process Quality Management and Improvement." A particular process improvement methodology, developed within AT&T Bell Laboratories. (AT&T Quality Steering Committee 1988b).

Process. The way in which anything is done. Anything a person does follows a process—often an unconscious process. Most men can tie a "regular" necktie without having to look at instructions, they have learned the necessary sequence of steps—the process. Asked to tie a bow tie, the same people would often be forced to turn to a set of pictorial instructions—a documented process.

Process Coach. A coach with skills in quality and in process management and improvement, specifically charged to help an individual, team, or organization improve their quality and their processes.

Process Gate. See **Quality Gate.**

Process Improvement. A specific activity intended to make some process more effective or efficient.

Process Management. The activity of consciously managing the processes used by an organization.

Process Management Team (PMT) A team (see below) formed for the specific purpose of documenting, monitoring, and improving a particular process.

Product Realization Process (PRP). The entire high-level process by which a new product is brought from concept to manufacture.

Product Team. A team (see below), usually multi-functional, formed to manage the realization of a new product.

Productivity Improvement. Improvement in some measure of an organization's productivity. Examples are shortened development intervals, lower development costs, and improved product quality.

Project Leader. The individual charged with the responsibility to carry out an identified project.

Project Team. A team (see below) specifically formed to carry out an identified project.

PRP. See **Product Realization Process**.

QIP. See **Quality Improvement Project.**

QIT. See **Quality Improvement Team**.

Quality. Traditionally a term used to reflect reliability or desirability of a product, or the efficiency or effectiveness of a service. More recent uses of the term extend to cover all attributes of a product, process, or service that affect its ability to meet the needs of customers and of the business offering those products, processes, or services.

Quality Gate. A point in a process where a set of established criteria are evaluated to determine if the next process can be initiated.

Quality Improvement Project (QIP). A project set up to development an improvement to solve an identified quality or process problem. A *QIP* is usually worked by a *QIT* (see below).

Quality Improvement Team (QIT). A team (see below) formed for the specific purpose of developing an improvement to solve an identified quality or process problem. Usually a QIT disbands when the sought-for improvement has been implemented. A QIT usually works on a specific QIP.

Quality System. The set of methods and systems that defines how an organization will achieve its quality goals, and enables it to do so.

R&D. Research and Development. Used loosely in many cases to refer to a company's development organization.

Readiness Assessment. A questionnaire, survey, or test designed to determine how ready a team is to begin work on a project. Typically a readiness assessment would look to see if the team was up to strength, if sufficient resources are available, if clear and firm requirements are available, if management has demonstrated commitment to the project, etc.

Recognition. Acknowledgment, usually public, of the achievements of an individual or team. Recognition may or may not be accompanied by a tangible reward, but usually includes some token such as a certificate.

Resources. The people, systems, and materials needed to accomplish a task.

Reward. Something tangible given to an individual or team as thanks for some achievement. May be money, a gift of value, extra time off, etc.

Robust. Capable of surviving or functioning over a wide range of conditions. A *robust design* will result in a product that functions as intended even when the components from which the product is manufactured vary over a wide range.

Role. The function played by each person in some activity. For example, typical roles in a team are "leader" and "member".

Root Cause. The *real*, underlying reason for some consequence—often a problem. In trying to improve a process problem, it is important to determine the root cause; otherwise the improvement effort may wastefully be applied to some symptom, and the problem may occur again.

SEI. See **Software Engineering Institute**.

Self-Directed Team. A team (see below) that, having been formed and given a goal by some external party, is left on its own to decide how the team members will work to achieve the goal. They establish their own roles, determine their own individual assignments, choose their own tools, solve their own problems, etc. Coaches and other resources are available to them, but serve only in advisory roles.

Sharing Rally. A gathering of people working toward a common goal, for the purposes of sharing experiences and lessons learned with everyone, and building enthusiasm.

Shewhart Cycle. Plan-Do-Check-Act cycle; also known as the *Deming Cycle*.

Shingo Prize. The *Shingo Prize for Excellence in Manufacturing* recognizes companies in the United States, Canada, and Mexico that demonstrate excellence in manufacturing leading to quality enhancement, productivity improvement, and customer satisfaction.

Simultaneous Engineering. Another term for the concepts embodied in *concurrent engineering*.

Slip Ratio. A means for evaluating the magnitude of a schedule slip in a normalized way: the actual slip divided by the scheduled interval for the slipped event.

Software Engineering Institute. Center associated with Carnegie Mellon University, sponsored by the United States Department of Defense, established to study and improve the design and engineering of software.

Storming. One of the stages in development of a team, during which the interpersonal issues of the team members are surfaced and dealt with. Successfully coming through this stage is crucial; if issues are not resolved here, they are likely to surface during some crisis and negatively affect the team's ability to perform.

Sub-process. A process that functions within a higher-level process. For example, a product realization process typically has detailed development processes, like software development, as sub-processes.

System. A general term used to refer to all of the elements of an operation including the process, doers, management, tools, environment, etc.

Team. A group of people working together in a planned, often structured way to achieve a task or job.

Team Building. Efforts by the members of a team to help them work together most effectively. These are usually shared, structured activities designed to help the team members get to know each other better, understand one another's skills and limitations, and find ways to work together easily and effectively.

Template. An outline or format for a document, schedule, checklist, etc., which makes developing application-specific documents, schedules, etc., easier.

Time-to-Market. The time between product conception and availability in the marketplace (also called *product realization interval*).

Tools. Artifacts or methodologies that an individual can use to help do a job. Examples include software applications, templates for developing checklists, the checklists themselves, and ways of working together like *brainstorming*.

UL. Underwriters Laboratories. A private organization that establishes safety standards for products, and then tests products for compliance with the applicable standards and certifies those that meet the standards. In the United States, many jurisdictions recognize UL standards and certification as adequate certification that products are safe.

Value Engineering. A methodology for analyzing a product proposal or design to ensure that all *necessary* features, functions, and characteristics are achieved, and that all features, functions, and characteristics that are included provide value, and are implemented in the most cost-effective way.

Values. Those underlying beliefs or principles that guide people's behavior. Examples of values adopted by a company might include diversity, respect for the individual, teamwork, integrity, and so on.

Variation. When applied to process metrics, *variation* refers to the fluctuation or change in the data. Most process metrics will have a natural variation due to the variation in materials, people, machines, and other factors such as environment. A certain amount of variation is considered normal, and the limits of this variation are called control limits.

Vendor. A third-party supplier of materials or services.

Vision. Commonly defined as a perception of a future state or condition through imagination of foresight. James A. Belasco, in his book *Teaching the Elephant to Dance*, formulates *vision* from a business perspective as "a statement of what you want your organization to be. It conveys a picture of where you want to go and how you want to get there. It is a simple-to-understand, inspirational, focusing statement" (Belasco 1990, 104).

World Class. Most strictly: proven to be competitive in a worldwide market, such as a product that outsells in many markets competitive products from vendors around the world. Commonly used to describe a process or product that is believed by its creators and users to be capable of achieving world-class competitive status. "The best there is."

360° Feedback. In personnel evaluation, the concept of seeking feed-back on an individual's performance from people representing all those with whom the individual interacts: coaches, peers, subordinates, team-mates, suppliers, customers, etc.

References

AT&T Writing Committee, Bonnie B. Small, Chair. 1956. *Statistical Quality Control Handbook*. Indianapolis, IN: AT&T Customer Information Center.

AT&T Quality Steering Committee. 1988a. *AT&T Quality Improvement Cycle*. Indianapolis, IN: AT&T Customer Information Center.

AT&T Quality Steering Committee. 1988b. *Process Quality Management & Improvement Guideliness*. Indianapolis, IN: AT&T Customer Information Center.

Bakerjian, R. 1993. *Tool and Manufacturing Engineers Handbook*. Vol. 7, *Continuous Improvement*. Dearborn, MI: Society of Manufacturing Engineers.

Belasco, James A. 1990. *Teaching the Elephant to Dance*. New York: Crown.

Bolman, Lee G., and Terrence E. Deal. 1992. "What Makes a Team Work?" *Organizational Dynamics* (Autumn): 34–44.

Brassard, M. 1989. *The Memory Jogger Plus*. Methuen, MA: Goal/QPC.

Bridges, William. 1991. *Managing Transitions: Making the Most of Change*. Reading, MA: Addison-Wesley.

Brown, Mark Graham. 1993. *Baldrige Award Winning Quality: How to Interpret the Malcolm Baldrige Award Criteria*. 3d ed. Milwaukee, WI: ASQC Quality Press.

Camp, Robert C. 1989. *Benchmarking: The Search for Industry Best Practices that Lead to Superior Performance*. Milwaukee, WI: ASQC Quality Press.

Carr, Clay. 1992. "Planning Priorities for Empowered Teams." *Journal of Business Strategy* (September/October): 43–47.

Christensen, L. B. 1977. *Experimental Methodology*. Boston: Allyn & Bacon.

Clausing, Don, and John R. Hauser. 1988. "The House of Quality." *Harvard Business Review* 66, no. 3 (May-June): 63–73.

Coulson-Thomas, Colin. 1992. "Strategic Vision or Strategic Con?: Rhetoric or Reality?" *Long Range Planning* 25, no. 1 (February): 81–89.

Deming, W. Edwards. 1986. *Out of the Crisis*. Cambridge, MA: Massachusetts Institute of Technology.

Dobyns, Lloyd, and Clare Crawford-Mason. 1991. *Quality or Else: The Revolution in World Business*. Boston: Houghton Mifflin.

Drake, Samantha. 1992. "Teaming Up." *Human Resource Executive* (August): 39–42.

Francis, Philip H. 1992. "Putting Quality into the R&D Process." *Research-Technology Management* (July-August): 16–23.

Gabor, Andrea. 1990. *The Man Who Discovered Quality.* New York: Random House.

Gatenby, David A. 1988. "Design for "X" (DFX) & CAE/CAD." *Proceedings of the Third International Conference on Product Design for Manufacture & Assembly, Newport, RI, 6–8 June*. Rochester, MI: Troy Conferences, Inc.

Gatenby, David A., and George Foo. 1990. "Design for X (DFX): Key to Competitive, Profitable Products." *AT&T Technical Journal* 69, no. 3 (May-June): 2–13.

Gatenby, David A., Paul M. Lee, Randall E. Howard, Kaveh Hushyar, Richard Layendecker, and John W. Wesner. 1994. "Concurrent Engineering: Enabler for Fast, High-Quality Product Realization". *AT&T Technical Journal* 73, no. 1 (January-February): 34–47.

Goldratt, Elihu M., and Jeff Cox. 1986. *The Goal: A Process of Ongoing Improvement*. Croton-on-Hudson, NY: North River Press.

Gomory, R. E. 1989. "From the 'Ladder of Science' To the Product Development Cycle." *Harvard Business Review* (November-December): 99–105.

Grady, R. B. 1989. "Dissecting Software Failures." *Hewlett-Packard Journal* (April): 57–63.

Hirschhorn, Larry. 1991. *Managing in the New Team Environment: Skills, Tools, and Methods*. Reading, MA: Addison-Wesley.

Humphrey, Watts S. 1989. *Managing the Software Process*. Reading, MA: Addison-Wesley.

Hunt, V. D. 1992. *Quality in America: How to Implement a Competitive Quality Program*. Homewood, IL: Business One Irwin.

Institute for Defense Analysis. 1988. *Concurrent Engineering*. Report R-338, December.

Ishikawa, K. 1982. *Guide to Quality Control*. Tokyo: Asian Productivity Organization.

Jones, C. L. 1985. "A Process-Integrated Approach to Defect Prevention." *IBM Systems Journal* 24, no. 2: 150–166.

Juran, Joseph M. 1960. "Pareto, Lorenz, Cournot, Bernoulli, Juran and Others." *Industrial Quality Control* 17, no. 4 (October): 25.

Juran, Joseph M. 1964. *Managerial Breakthrough*. NY: McGraw-Hill.

Kano, Noriaki, Nobuhiko Seraku, Fumio Takahashi, and Shinichi Tsuji. 1984. "Attractive Quality and Must-Be Quality." *Quality* 14, no. 2:39–48.

Katzenbach, Jon R., and Douglas K. Smith. 1993. "The Discipline of Teams." *Harvard Business Review* (March-April): 111–120.

Kawasaki, Guy. 1990. *The Macintosh Way*. Glenview, IL: Scott, Foresman.

Kearns, David T., and David A. Nadler. 1992. *Prophets in the Dark: How Xerox Reinvented Itself and Beat Back the Japanese*. New York: Harper Business.

King, Bob. 1989. *Hoshin Planning: The Developmental Approach*. Methuen, MA: GOAL/QPC.

Kroeger, Otto, and Janet M. Thuesen. 1988. *Type Talk*. New York: Delta Books.

Kume, H. 1985. *Statistical Methods for Quality Improvement*. Tokyo: Association for Overseas Technical Scholarship.

LaMarsh, Jeanenne. In Press. *Changing the Way We Change*. Reading, MA: Addison-Wesley.

Machiavelli, Niccolò. 1947. *The Prince*. Translated and edited by Thomas C. Bergin. New York: Appleton-Century-Crofts.

McArdle, Kevin J. 1993. "Recognition: A Powerful Tool for Continuous Improvement." *Quality Digest* (July): 33–37.

McCusker, Joseph. 1992. "Simultaneous Engineering at Cadillac." Presented at *SME Clinic, How to Implement Concurrent Engineering*, 5–6 May, Southfield, Michigan.

Mizuno, S. [1979] 1988. *Management for Quality Improvement: The Seven New QC Tools*. Cambridge, MA: Productivity Press.

Myers, M. Scott. 1990. "The People of Milliken." *Target* (Fall): 16–19. (Published by the Association for Manufacturing Excellence.)

Osborn, Alex F. 1963. *Applied Imagination*. 3rd ed. New York: Charles Scribner & Sons.

Patterson, Marvin L. 1993. *Accelerating Innovation*. New York: Van Nostrand Reinhold.

Peters, Thomas J., and Robert H. Waterman, Jr. 1982. *In Search of Excellence*. New York: Harper & Row.

Peters, Thomas J., and Nancy Austin. 1986. *A Passion for Excellence*. New York: Warner Books.

Putnam, A. O. 1985. "A Redesign for Engineering." *Harvard Business Review* (May-June): 139–144.

Scherkenbach, William W. 1986. *The Deming Route to Quality and Productivity*. Washington, DC: CEE Press, George Washington University.

Shewhart, Walter A. 1986. *Statistical Method from the Viewpoint of Quality Control*. New York: Dover Publications. (Originally published in 1939 by the Graduate School of the U.S. Department of Agriculture, Washington, DC.)

SME. 1992. From presentation at clinic on *How to Implement Concurrent Engineering*. 5–6 May, Southfield, Michigan.

Smith, Bryan. 1989. "Vision: A Time to Take Stock." *Business Quarterly* 54 (Autumn): 80–84.

Spendolini, Michael J. 1992. *The Benchmarking Book*. New York: AMACOM, a division of American Management Association.

Stahl, Susan S. 1992. *The Empowering Supervisor*. King of Prussia, PA: Organization Design and Development.

Stanzione, Daniel C. 1992. Paper presented at AT&T Bell Laboratories' 1992 *Quality Symposium*. 14 September, Holmdel, NJ.

Taguchi, Genichi. 1993. *On Robust Technology Development*. New York: ASME Press.

Torres, Crescencio, and Jerry Spiegel. 1990. *Self-Directed Work Teams: A Primer*. San Diego, University Associates.

Tuckman, Bruce W. 1965. "Developmental Sequence in Small Groups." *Psychological Bulletin* 63, no. 6: 384–399.

Watson, Gregory H. 1992. *The Benchmarking Workbook: Adapting Best Practices for Performance Improvement*. Cambridge, MA: Productivity Press.

Webb, David. 1992. "HP and Compaq Make Empowerment Work." *Electronic Business* (December): 54–57.

Wellins, Richard S. 1992. "Building a Self-Directed Work Team." *Training & Development* (December): 24–28.

Wesner, John W. 1991. "DFM at a Distance." *Concurrent Engineering: Issues, Technology, and Practice* 1, no. 1 (January-February): 5–13.

Yeatts, Dale E., Leslie Stanley-Stevens, and Paul Ruggiero. 1992. "The Self-Managed Work Team: Reasons for Failure and Potential Solutions." *QPM* 9, no. 4:26–32. (Published by the Virginia Productivity Center, Virginia Polytechnic Institute & State University.)

Index

360° feedback, 124-25, 154
80/20 rule. *see* Pareto principle

Achievers' Club, 260
Affinity Diagram, 84, 85, 87-88, 226
  effective group size, 88
  rules for, 85
affirmation, 260
after-the-fact reviews, 108
alignment, 163, 220, 223
  horizontal, 223
  metrics. *see* metrics
  with business objectives, 182
American Productivity and Quality Center, 209, 214
American Society of Mechanical Engineers, 209
annual reviews, 254
ASME. *see* American Society of Mechanical Engineers
assessment, 167
  concurrent engineering success factors, 134
  external. *see* external assessment
  internal. *see* internal assessment
  readiness. *see* readiness assessment
  quality programs, 251
  self. *see* self-assessment
architecture, process. *see* process architecture
associate satisfaction, 137, 194, 196
  see also employee satisfaction
AT&T Bell Laboratories, 3, 212, 264
AT&T Value Statement, 29
awards, 261, 263

Belasco, James A., 20, 184, 186
benchmarking, 23, 24, 69-70, 179, 210, 214, 218, 241
  areas to investigate, 179-80
  companies to benchmark with, 180-82
  data, 13, 234
  methods for gathering information, 180
  process for, 179
best current practices, 13, 69
best-in-class, 183
"best of the best" process, 13, 214
block diagrams, 42, 43, 48
bonus, 262
Boothroyd and Dewhurst, 108
breakthrough improvements, 35, 61
brainstorming, 82, 84, 87, 226
  rules for, 84-85
British Institute Of Management, 21
Brooktree Corporation, 139
budget, 220, 237
  for process, 217-18
build enthusiasm, 228
business case, 220

Cadillac Motor Car Corp., 139, 212
Camp, Robert C., 70, 181
cards, business, 237
cards, pocket, 229, 236
cascade, 227-28, 232
cause and effect diagram. *see* root cause diagram
celebrating, 197, 259
celebrations, 261, 267-68
change
  see also organizational change
  agents, 218, 221
  managing, 163, 195, 208
changing roles, 135
charging case, 220
checklists. *see* process checklists
checkout line, example, 34
check sheets, 94, 95
Class A metric, 49-50
coach
  process. *see* process coach
  team. *see* team coach
coaching skills, 143
co-location, 109
comfort zones, 261
commitment at the top, 19, 25-26
common bond. *see* AT&T Value Statement
communication protocol, 118
communication, concurrent engineering, 114, 117
communications, 139, 158
compensation, 228
competitor benchmarking. *see* benchmarking
competitor performance level. *see* performance level
component lead time, risk management, 129-30
Concurrent Engineering, 105, 113, 126, 136-37, 148, 177, 231, 235, 245, 251, 252
  benefits of, 111
  communication, 117
  definition, 105
  evolution towards, 113
  getting started, 116
  incremental information exchange, 126-27
  part time vs. full time teams, 115-16
  process
    design of, 125, 131
    objectives of, 125
  quick assessment, 134
  risk management, 129-30
  story, 105
  success factors, 114
  team formation, 114-15
  teams, 113-14
conferences, 209
confidence level. *see* risk management
CONRAIL (Consolidated Rail Corporation), 236-37
consultants, 181, 226
continuous improvement, 168, 212, 249, 255
  vs. re-engineering, 62
continuous process improvement model, 40-41
control charts, 94
core team, 155, 218, 221, 235-36, 238, 269
cost reduction, 137, 217
cultural strain, 163
culture, 155, 267
  new, 224, 261
  traditional, 4
Customer Action Team, 23, 24
customer
  see also process, customers
  and supplier, when same, 40
  employees as, 172

focus for re-engineering, 66
expectations, 4
identification, 171-72
key drivers for change, 19, 22, 24
needs, prioritizing, 174, 175
perceptions, 13
process customers, 38
requirements and needs, 36, 46, 48
sample list for product development, 173
satisfaction, 22, 194, 203, 205
survey, 176
voiced vs. unvoiced needs, 48, 68
customer-supplier model, 38
customers for life, 22
cycle time, 4

data, 199, 208, 215, 223, 242
data driven decisions, 94
Deming, W. Edwards, 18, 22, 24, 154, 169, 209, 240
Prize, 82, 212
deployment, 203
goals, 188
plan, 77
program, re-engineering, 76
re-engineered processes, 77
design
for manufacturability, 107-9, 232, 246
reviews. *see* after-the-fact reviews
for 'X', 108, 112, 232
of experiments, 226
transfer, 216
design and manufacturing relationship, 105
develop your program, 193
development
environment, 4
intervals, 108
team, 149
DFM. *see* Design for Manufacturability
at a distance, 109
guidelines, 108-9
DFX. *see* design for 'X'
diagnostic measures. *see* measures
digraphs, interrelationship. *see* interrelationship digraphs
dinner for two, 265, 266
documentation
process, 41, 152, 203, 240
re-engineered processes, 74-75
documented process, 152
documenting the processes. *see* documentation, process

electronic mail, 117
email. *see* electronic mail
protocol. *see* communication protocol
employee
involvement, empowerment, 19, 21, 24
satisfaction, organizational enabler, 27-28
suggestion programs, 23, 24
satisfaction, 27-28, 172
see also associate satisfaction
employees as customers, 172
empowered
employees, 19, 137
teams, 139, 141, 155
see also self-directed teams
empowerment, 19, 21, 24, 137, 141, 155, 198
relation to middle managers, 26
enabling objectives, 75
entry and exit criteria, 55
environment for quality, 4, 137, 142, 148, 162, 163, 224-25, 233, 261, 262
evangelizing, 221
experiments, design of. *see* design of experiments
external
assessment, 9, 12
focus, 68, 69

FCC (Federal Communications Commission), 235
Part 15, 217
Federal Express, 24, 29
feedback, 203
fishbone diagram. *see* root cause diagram
Five-Minute Self-Assessment, 9, 10, 195-96
Florida Power and Light, 82, 212
flowcharts, 42, 43, 48
focus groups, 175
forming (stage in team development), 164, 239
Francis, Philip, 172
frustration, 210
full-time participation, 148
full-stream considerations, CE process, 133
functional
manager, 146-48, 151, 154, 157, 161, 165, 168
see also middle managers
processes, 148, 161, 240, 245
funding, 207, 217-18, 232
obtaining, 207, 208
future state, 195, 196
see also vision

gaining support, 204
gap, 13, 199, 202
see also performance gap
Gatenby, David A., 108
gifts, 154, 260, 264, 267
mementos, 268
goal
deployment, 188
setting, 169
table, example for policy deployment, 189
goals, 142, 160, 195, 203, 219, 231, 236, 259
clear, 165
for product development, 187
golden thread, 188
grass roots effort, 232

habits, 261
challenging, 261
hats, 157, 268
health matrix. *see* process health matrix
heroism, 4

Hewlett-Packard, 139, 212
high performance work teams, 23
Hoshin planning, 169, 226
  see also Policy Deployment

identify opportunities and implement solutions, 59
ideological change, 163
IEEE. *see* Institute of Electrical and Electronic Engineers
implementation, 193
incremental information exchange, 126
infrastructure building, TQM, 255
inputs, process. *see* process, inputs
Institute of Electrical and Electronic Engineers, 209
internal
  assessment, 9, 243
  focus, 34
interpersonal skills, 151, 159
Interrelationship Digraph, 87, 88-89, 90, 96, 109, 226
  rules for constructing, 88-89
intervals, 14, 137, 138, 139, 198, 214, 215, 247
  impact on process management, 36
  manufacturing readiness, 112
  printed wiring board development, 111
interviews, 174
invention, 244, 247
Ishikawa, K., 87
Ishikawa diagram. *see* root cause diagram
ISO 9001, 203

John Deere Co., 212
journey, 3, 5, 9
Juran, Joseph, 92, 209

Kume, Hitoshi, 36

LaMarsh, Jeanenne, 195
language, common, 225
leader. *see* team leader
leadership, 19, 25, 145, 146, 149, 151, 161, 203
  qualities of good leaders, 25
  skills, 162
learning curve, 76
library, process, 243
logo, 154, 186, 267, 268
luncheons, 268

Machiavelli, 194
magnet, 267
Malcolm Baldrige National Quality Award, 11, 139, 199, 268
management support, 55
managers, role in CE, 122
managing
  change. *see* change, managing
  quality programs, 249
manufacturing, 232, 235
  process design, 131
  readiness interval, 126
mapping quality programs to quality ideals, 23
market share, 194
matrix, mapping of quality programs, 250
maturity, 152
measures
  see also metrics
  diagnostic, 47
  lack of, 33
  performance, 46
  process, 38, 40
meetings
  all-hands, 190
  CE teams, 120
mementos, 261, 268
mentor, 162
metrics, 57, 187, 203
  alignment of, 44
  catch-22 for product development, 47
  criteria for good, 50
  establishing, 46, 48
  focus on results, 53
  performance assessment, 23, 24
  quality ideal, 24
  vertical and horizontal alignment, 52
middle managers, 163
  support, 19, 26
Milliken Company, 22, 24, 179, 268, 269
mission, 203, 219, 231, 259
Motorola, 18, 24, 179
motto, 241
multi-function
  teams, 136, 138, 139, 140
    design, 136
multi-voting, 84, 86, 92
  rules for, 86
Myers-Briggs Type Indicator, 260

National Technological University, 219
newsletter, 241, 264
non-cooperative manager, 163
norming (stage in team development), 164, 239
NTU. *see* National Technological University

objectives, 142, 203
  alignment. *see* golden thread
  business, 182
  shared, for CE teams, 114, 121, 126
objective setting. *see* goal setting
one cycle, one team, 268
OOPS, 216-17
opposition, 234
organizational
  change, 30, 79, 256
  differences, impact on CE, 122
  enablers, 25, 135, 193, 207, 231, 259
  skills, 159
Osborn, Alex F., 85
outputs, process, 38
outside
  consultants. *see* consultants
  direction, 151
over-the-wall design transfer, 106

pair-wise comparison. *see* interrelationship digraph
paradigm shift, 5, 194
paraphernalia. *see* gifts
Pareto
  chart, 82, 83, 92-93, 95, 98, 100, 101
    rules for, 93
  principle, 92
part-time participation, 148
participation
  full-time, 148
  part-time, 148
PDCA cycle, 255
peer
  awards, 263
  recognition, 266
people
  role of, 4
  tools, 135
People-Service-Profit philosophy, 29
performance
  evaluation, 153, 203
  gap, 181, 182
  level, against competitors, 179
  measures, 46
  review, 57, 153, 203
performance, tracking, 120
performing (stage in team development), 164
Peters, Tom, 211
pilot, 205, 215, 235, 238
  see also trials
Plan-Do-Check-Act cycle, 169, 190, 250, 255
plaques, 229
platforming, 247
PMT. *see* process management team
pocket cards, 229, 236
Policy Deployment, 20, 24, 169-70, 251, 252
PQMI, 34, 226, 242
Pre-Project Readiness Assessment, 243, 244
President's Quality Award, 264
prioritization
  matrices. *see* prioritization worksheets
  of causes and solutions, 92
  worksheets, 92, 93, 96
    rules for, 93
problem solving, 59
problem solving process, 96
process
  adherence to, 54
  architecture, 238
  basics, 38
  boundaries, 38
  checklists, 44
  coach, 145, 147, 156, 160, 168, 239, 243
  competitors, 13
  customers, 38
  definition, 38
  deployment assessment, 253
  diagrams. *see* block diagrams and flowcharts
  documentation, 41-44, 203
  effective deployment of, 54
  focus, 136
    on How Of Doing Work, 22, 24, 26
  gates. *see* quality gates
  health matrix, 62
  hierarchy, 44
  improvement, 23, 33
    methodology, 33, 236
  inputs, 39
  library, 243
  management, 137, 203, 210, 219, 231
  management methodology, 231, 232, 233
  management team, 140, 148, 149, 224, 239, 241, 245, 264
  measuring performance, 57
  metrics. *see* measure and metrics
  outputs, 39
  owner, 146, 148, 157
  purpose, 38
  steps, 38
  suppliers. *see* suppliers
  team, 155
  team leader, 147, 157
Process Health Matrix, 200
process management, 23
  background, 35
  benefits of, 35
  building blocks, 36
  differences from manufacturing, 36
  for product development, 36
  in manufacturing, 35
  in R&D, 37
  methodology, 33
  model for, 41
  structure, 61
  success indicators, 60
  teams, 23
process re-engineering, 61
  choosing, 61
  customer needs, 68
  design principles, 70
  on-going management, 78
  principles, 66
  process principles, 68, 70
  prospectus, 67
  step 1. Concept Development, 67
  step 2. Planning and Definition, 72
  step 3. Development, 74
  step 4. Verification, 76
  step 5. Deployment, 77
  team formation, 67
  tool requirements, 73
  vs. process management, 61
processes, 152
product
  cost, 198
  quality, 138
  teams, 137
product development, 232
  team, 140, 141
product realization process, 111, 141, 196, 198, 214, 245
productivity, 17, 18, 35, 38, 39, 40, 137
  improvement, 3, 137

profound knowledge, 240
programs. *see* quality programs
project
  leader, 137, 147, 157
  management tools, 153
  retrospectives, 58-59
  team, 137
prospectus. *see* process re-engineering
prototype cycles, 139
PRP. *see* product realization process
psychological type, 260
public recognition, 266
Putnam, Arnold, 106

QFD. *see* Quality Function Deployment
QIP. *see* quality improvement project
QIS. *see* Quality Improvement Story
QIT. *see* quality improvement team
QPD. *see* Quality Policy Deployment
qualities of a good leader. *see* leadership
quality
  cascade, 227-28
  circles, 28
  gates, 44, 55-56, 58, 71, 75
    senior manager involvement, 56
  overview of concepts, 17
Quality Function Deployment, 24, 226
quality improvement
  project, 141, 211, 216, 245
  teams, 23, 141, 149, 245
Quality Improvement Story, 82
Quality Policy Deployment, 24, 236
  see also Policy Deployment
quality system, 17, 170
  design of, 249, 255
  long term planning, 252
  matrix of quality programs, 251

radar chart, process deployment results, 254
raises, 262
readiness assessment, 152
recognition, 153, 155, 228, 259
re-engineering. *see* process re-engineering
repeated questioning, 87, 91, 100
  rules for, 91
report card, responsiveness, 119
requirements, 216
resource
  center, 243
  requirements for re-engineering, 63
resources, 207, 220, 231, 237, 243
results, 211, 234
reward
  structure, 30
  system, 106, 256
rewards, 153, 259, 261
  mementos, 261
  monetary, 261, 262, 263
    team, 262
  public recognition, 266
  shared in CE teams, 114, 121
risk management, 129
robust design, 217, 226
role
  model, 221
  reversal, 49
roles, 144, 148, 156
  new, 161
  responsibilities, 73
    of process users, 45
root cause, 59, 68, 71, 87, 213, 216, 217
  analysis, 82, 83, 87, 91, 95, 97, 101
  categorizing, 87
root cause diagram, 83, 87
  rules for, 87
run charts, 94
  rules for, 94

satisfaction, associate. *see* associate satisfaction
schedules, 153
scope, process. *see* process boundaries
self-assessment, 10, 13
self-directed. *see* empowerment
  teams, 137, 141, 148, 152, 155, 159, 162, 164, 203, 239, 262
  work teams. *see* high performance work teams
seminars, 224
senior managers, 79, 171
  involvement in quality gates, 56
Seven Basic Tools. *see* Seven QC Tools
Seven Management Tools, 95, 227
Seven New Tools. *see* Seven Management Tools
Seven QC Tools, 95
sharing rally, 213, 269
Shewhart, Walter A., 18, 24, 169, 236
Shingo Prize, 12
shirts, 268
simultaneous engineering, 212
  see also Concurrent Engineering
skilled staff, 202, 210
slip ratio, 215
slogan, 154, 241
SME (Society of Manufacturing Engineers), 209
Smith, Bryan, 21
software, 218
Software Engineering Institute, 199
software process maturity, 199
solutions, to process problems, 59
SPC. *see* statistical process control
Spendolini, Michael J., 70, 181
split up the work, 237
sponsor, 151
state
  future, 195, 196
  target, 197
statistical process control, 22, 24, 35
storming (stage in team development), 164, 239
Strategic Quality Ideals, 19-22, 23-24, 25, 135, 193, 207, 231, 259
sub-process, definition of, 44
success factors for concurrent engineering, 113
success stories, 210, 213

suggestion programs, 23, 24
supervisors, role in CE, 122
suppliers
  feedback, 54
  process, 38
  requirements on, 53
support structure, 207, 231, 232
surveys, 174
  customer, 174
  manufacturing, 177
  product management, 176
  responsiveness, 119
symptom analysis, 87
system, 36

Taguchi, Genichi, 209, 222
target state, 197
  see also future state and vision
team
  award, 155, 264, 267
  building, 142, 165, 219
  coach, 145, 147, 150, 156, 160, 164, 168
  formation, 114, 148
  leader, 145, 147, 149, 150, 151, 154, 156, 159, 266
  management model, 46
  members, 135, 145, 147, 148, 151, 156, 158
  membership, 142
  pictures, 266
  size, CE teams, 114
team-based structures, 21
teams, 135, 136, 138, 140, 147, 225, 228, 232, 262
  empowered. *see* empowered teams
  multi-function. *see* multi-function teams
  self-directed. *see* self-directed teams
teamwork, 137, 138-39, 144, 158
technical experience, 161
templates, 153
tennis ball exercise, 62
time-to-market, 46, 56
  see also interval
tool architecture, 73
tools, 81, 234
  common, 207, 225
  for teams, 152
  in concurrent engineering, 133
top management support, 146
total quality management, 249, 255
TQM. *see* Total Quality Management
tradition, 155
training, 55, 74, 157, 162, 164, 202, 203, 210, 218, 219, 226, 227, 232, 256
trials, for re-engineered processes, 76

understanding where you are, 9, 194, 197
understanding where you want to go, 13, 194
Underwriters Laboratories, 235
using teams, 198, 225, 229, 234, 238-39
using this book, 6, 11

value
  engineering, 143
  statement, example, 122
  system, 110
values, 29, 143, 163, 167, 219, 247
  see also AT&T Value Statement
  shared in CE teams, 114, 121
variation, 35
vendor partnership programs, 23
  see also suppliers
verification, 76
vertically aligned process structure, 52
video teleconferences, 117
vision, 19, 20, 21, 24, 196, 203, 219, 229, 231, 236, 259, 261
  creating for product development, 184
  example, 185
  in policy deployment process, 169
  process re-engineering, 67
voice mail, 117

walk the talk, 161
warranty experience, 215
Watson, Gregory H., 70, 181
Wesner, John W., 109
what-how-who, 4
working smarter, 5

Xerox Corporation, 212